Agricultural engineering in development

The organization and management of replacement parts for agricultural machinery
Volume 2

Compiled and written by
Roger Limbrey

Edited and prepared for publication by
Stephen Dembner

FAO AGRICULTURAL SERVICES BULLETIN

72/2

Deutsche Gesellschaft für Technische Zusammenarbeit

FOOD AND AGRICULTURE ORGANIZATION OF THE UNITED NATIONS
Rome, 1988

The designations employed and the presentation of material in this publication do not imply the expression of any opinion whatsoever on the part of the Food and Agriculture Organization of the United Nations concerning the legal status of any country, territory, city or area or of its authorities, or concerning the delimitation of its frontiers or boundaries.

ISBN 92-5-102700-5

All rights reserved. No part of this publication may be reproduced, stored in a retrieval system, or transmitted in any form or by any means, electronic, mechanical, photocopying or otherwise, without the prior permission of the copyright owner. Applications for such permission, with a statement of the purpose and extent of the reproduction, should be addressed to the Director, Publications Division, Food and Agriculture Organization of the United Nations, Via delle Terme di Caracalla, 00100 Rome, Italy.

© FAO 1988

ACKNOWLEDGEMENTS

FAO acknowledges with thanks the assistance provided by the Tractor Division of Messrs Klockner-Humboldt-Deutz and the Caterpillar Tractor Company in allowing the author to delve into the insights of their policies and organization for the provision of replacement parts. In particular, thanks are extended to replacement parts distributors in Mozambique, Niger and Zambia who permitted photographs to be taken in their premises.

FOREWORD

Since 1945 there has been a rapid increase in agricultural mechanization. In the developed countries, the tractor has largely displaced draught animals as a source of power and a steady decline in the number of people working in agriculture has provided a stimulus to increased mechanization and productivity. This has led to a corresponding increase in competition for the farmer's business and to increased emphasis on aftersales service. In the developing countries, there have been equally significant developments in agriculture. The need for additional food production to feed rapidly increasing populations has led to emphasis on higher productivity through greater use of inputs in general and mechanization in particular. In many instances, mechanization has been introduced without the infrastructure necessary to support it. In most developing countries, mechanization is dependent on imported items. These and other factors have complicated the establishment of efficient aftersales service.

This publication is concerned with one aspect of aftersales service - that of replacement parts for agricultural machinery. Some farmers have access to aftersales service which enables them to maintain and operate their machinery without undue inconvenience throughout its working life. Unfortunately, there are many instances where machines are inoperable, sometimes at critical periods in the production cycle, because either a replacement part or the skill to fit it is unavailable. There are other instances of less immediate but nonetheless unsatisfactory consequences - where parts are prohibitively expensive; where the level of operator or service/repair skill is too low; where machinery is incorrectly used or mismatched or where soil and climatic conditions cause greater than anticipated replacement and repair problems.

The need for efficient parts provision within the wider function of aftersales service in the developing countries is therefore clear. No single interest can achieve this:

- Governments have an overall responsibility for policy and implementation;
- Manufacturers and distributors have technical and commercial responsibilities;
- Farmers have operational and maintenance reponsibilities;
- Multilateral and bilateral aid agencies have roles to play in conjunction with other interests and particular responsibilities on parts and aftersales service requirements in projects they undertake or promote;
- International organizatons, including agencies in the United Nations system, have responsibilities in the provision of realistic and appropriate advice on all aspects of agricultural mechanization.

FAO and Deutsche Gesellschaft für Technische Zusammenarbeit (GTZ), the official technical assistance organization of the Federal Republic of Germany have jointly contributed to make this publication possible. This clearly demonstrates common goals and the valuable cooperaion which can be established between international and bilateral assistance organizations.

R.C. Gifford
Chief
Agricultural Engineering Service
Agricultural Services Division
FAO
Rome

J. Zaske
Head
Division of Agricultural Mechanization,
Agro-Industries, Technical Planning
GTZ
Eschborn, Federal Republic of Germany

Table of Contents

		Page
I.	Introduction	1
II.	Administrative Procedures	4
2.1	General concepts	4
2.2	Functions and duties	6
2.3	Order administation	6
2.4	Payment terms	8
2.5	Customs clearance	12
2.6	Emergency orders	15
2.7	Back orders	16
2.8	Stock administration	17
2.9	Stock checks	21
2.10	Removal of parts	24
2.11	Sales administration	26
2.12	Clerical functions	31
2.13	Inventory management techniques	34
III.	Management	42
3.1	Planning objectives	42
3.2	Continuity of supply	43
3.3	Planning details	45
3.4	Responsibility and authority	45
3.5	Control reporting	46
3.6	Stock level control	46
3.7	Movement reporting	48
3.8	Service level reporting	50
3.9	Back order reporting	53
3.10	Order control	55
3.11	Reasons for budgeting	57
3.12	Budget headings – income	58
3.13	Pricing	61
3.14	Budget headings – expenditure	64
3.15	Salaries	65
3.16	Other expenses	66
3.17	Purchase control	67

		Page
IV.	Staff	69
	4.1 Organization	69
	4.2 Staff responsibilities	71
	4.3 Parts manager	73
	4.4 Warehouse supervisor	78
	4.5 Clerical supervisor	81
	4.6 Sales supervisor	84
	4.7 Warehouse chargehand	88
	4.8 Warehouse staff	91
	4.9 Records clerk	93
	4.10 Order clerk	96
	4.11 Accounts clerk (for organizations where accounting is not centralized)	98
	4.12 Counter clerk (sales department: retail or workshop)	100
	4.13 Other staff	104
	4.14 Perpetual inventory supervisor	104
	4.15 Staff selection	107
	4.16 Training	110
	4.17 Staff control	116
V.	The Store: Facility and Organization	119
	5.1 Design and construction	121
	5.2 Layout	125
	5.3 Shelving and storage equipment	129
	5.4 Mechanical handling equipment	132
	5.5 Office equipment	133
	5.6 Stock card index	135
	5.7 Card design	137
VI.	Stock Planning	144
	6.1 Service level and investment	144
	6.2 Stock budget	145
	6.3 Types of parts: maintenance parts	148
	6.4 Consumables	151
	6.5 Hardware	155
	6.6 Wearing parts	160
	6.7 Service exchange units	164
	6.8 Non-wearing parts	169
	6.9 Suggested stock lists	170
	6.10 Stock in relation to machine type	170
	6.11 Motor vehicles	171
	6.12 Tractors	173

		Page
6.13	Cultivation implements	175
6.14	Planting equipment	176
6.15	Fertilizer and manure spreaders (including combine drills)	177
6.16	Sprayers	177
6.17	Forage equipment	177
6.18	Combine harvesters	178
6.19	Pumps and irrigation equipment	179
6.20	Grain handling and feed preparation machinery	182
6.21	Construction equipment and forklifts	183
Appendix A.	Ordering, Invoicing and Stocking Forms	185
Appendix B.	Administrative Forms	199
Appendix C.	Suggested Training Programme	203

I. INTRODUCTION

This is Volume 2 of a two-part publication, <u>The Organization and Management of Replacement Parts for Agricultural Machinery</u>. This volume has been prepared especially for personnel directly involved in the day-to-day operations of a replacement parts store. It provides detailed instructions regarding the techniques that can be used, the procedures and methods to be installed, and the controls to be applied to ensure economic, efficient operation which satisfies customer demand. This volume may be used as a procedural manual by the parts manager, or as a training guide.

Chapter 2 covers the administrative procedures relative to a replacement parts system. It begins with a discussion of the inter-relationship between the supply and demand cycles of parts supply. The chapter then details the administrative functions required on a regular basis to keep the two cycles in motion: ordering from the manufacturer; payment to manufacturers; receiving parts; handling incoming manufacturers' invoices; storing parts; receiving customer orders; dispatching parts to customers; writing customer invoices; receiving customer payments.

Chapter 3 deals with management responsibilities. The task of the parts manager is to keep the entire replacement parts operation moving smoothly. This means planning ahead, anticipating demand and reacting quickly and correctly to problems or unforeseen circumstances. The chapter covers the mechanisms needed to permit the parts manager to maintain control over the entire operation. Also covered are pricing and budgeting guidelines.

Chapter 4 covers the appropriate methods for staffing a replacement parts department. It is not enough to simply hire personnel or even to make provisions for training them. Staffing must be a carefully-planned and executed process, which ensures that the right person is hired for each job. The chapter begins with an organizational chart which helps to identify the posts to be filled. It then provides a summary of responsibilities and detailed job description for each position in a replacement parts operation, from the parts manager to the stocking clerks. The job descriptions detail the tasks which fall to each position, the authority of each post and the relationship of the post in question to the rest of the organization. In addition, a profile is provided which describes the necessary qualifications for each post. This chapter can be used both as a hiring guide and a tool to organize and evaluate staff. The chapter concludes with suggestions for an ongoing staff training programme.

Chapter 5 is a guide to the organization of the physical premises of the parts store and the stock it contains. Suggestions are given for store layout and for choice of storage facilities. Office layout is also covered. In addition, this chapter details the design and implementation of the stock card index, the information centre of the parts department.

Chapter 6 covers stock planning, i.e. the key decisions regarding what to stock and in what quantities. In the first part of the chapter, stocks are divided into five broad categories: maintenance parts; consumables; hardware, wearing parts; and non-wearing parts. Guidelines are provided to enable the parts manager to make the appropriate stocking decisions for each of these categories. In the second half of the chapter, stocking decisions are related to the various type of machine for which parts are required. Categories include on-road motor vehicles; agricultural tractors; cultivation implements; planting equipment; fertilizer and manure spreaders (including combine drills); sprayers; forage equipment; combine harvesters; pumps and irrigation equipment; grain handling and feed preparation machinery; and construction equipment and forklifts.

Appendix A presents forms which may be useful in organizing the stock ordering and management process. Appendix B consists of forms designed to facilitate personnel management. In both cases, many of the forms contained in these appendices are also presented in the appropriate sections of the main body of the text. They are reproduced in the appendices to facilitate ease of reference. Appendix C offers a suggested staff training programme.

The companion to this volume, Volume I is directed primarily at policy-makers and planners, those who determine the strategies, formulate plans, organize credit facilities and arrange import controls for the sale of replacement parts. It details the infrastructural requirements for the establishment and continuous operation of an efficient replacement parts operation; it is designed to equip decision-makers with sufficient knowledge to make adequate provision for replacement parts operations relating to the development of agricultural mechanization as a whole.

Although the two volumes of this publication are intended for people at different levels in the organization and management of replacement parts, it is recommended, nonetheless, that readers familiarize themselves with both volumes, for a comprehensive overview of a replacement parts system.

II. ADMINISTRATIVE PROCEDURES

2.1 General concepts

   The parts department is a vital link in the chain between manufacturer and user. It receives stocks, holds them and then dispatches them to the customer. Another major role is resolving potential conflicts of interest between the user who would prefer to invest nothing in parts stock, and the manufacturer who would like to maximize profits. The parts department, therefore, must decide what the user will need and when, and how that need can best be satisfied at minimum cost. At the same time, the parts department must convince the manufacturer and the customer that their interests are fully protected.

   The supply and demand cycles of a parts supply oper-ation interact at the parts department where orders and payment are sent to the manufacturer and parts are stocked; and where customers can order and buy parts (a diagram of the two interacting cycles is shown on page 5). The key to a successful operation is the simultaneous function of both cycles. A bottleneck in one can disrupt flow in the other.

   Another important factor is that the time period over which the two cycles operate varies for each part and each supply source. This is important because the point of stocking parts is to eliminate the user's waiting period while accounting for supply lead time. The sales cycle is geared to parts delivery on an as-needed basis. The supply cycle is designed to replenish stock just before it runs out so that the sales cycle is never interrupted. Since lead time may be lengthy, the skill in parts management lies largely in gauging the rate at which parts will move and ordering carefully calculated quantities at suitable intervals to ensure that stock does not run out.

While the sequence of payment, invoicing and dispatch may differ according to pre-arranged terms of payment, the two interlocking cycles remain constant. Each step in both cycles is an administrative function.

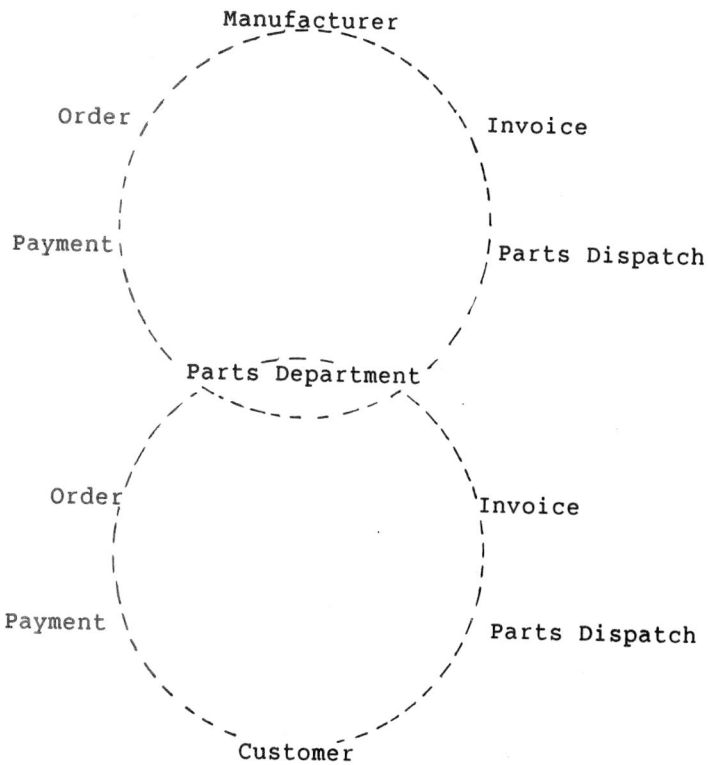

The Business Cycle

2.2 Function and duties

Certain functions must be performed on a regular basis to keep the two cycles in motion. The degree of specialization depends on the size and workload of the department. For example, in a very small store, one person may carry out all tasks. In a large store, more people performing specific functions will be required to carry the workload. The basic functions are:

- ordering from the manufacturer;

- payment to manufacturers;

- receiving parts;

- handling incoming manufacturer's invoices;

- storing parts;

- receiving customer orders;

- dispatching parts to the customer;

- writing customer invoices;

- receiving customer payments.

2.3 Order administration

Parts ordering is a clerical function that is very important to maintain because if parts are not ordered customers will be unable to obtain them as needed and the objectives of the parts supply operation will not be fulfilled. Ordering is not simple, since it involves the compilation of much detailed information. Unless strict discipline is imposed, the flow of orders may be interrupted and stocks may drop to dangerously low levels - a costly situation to remedy. Customers can become dissatisfied and the cost of emergency orders may have to be absorbed by the parts department.

There are two basic types of order, the normal stock order and the emergency order. The greater the precision in placing normal stock orders, the less will be the need for emergency orders; nevertheless accidents will always necessitate a small proportion of emergency orders.

In a large organization the two types of order may be handled separately, but normally they are combined as two functions under one person.

The information required for stock orders comes entirely from the stock control system, normally a card index. The card index holds all information relating to the parts, and access to it is necessary for various functions. In larger stores, there is a tendency to assign information entry and extraction duties to the same person, or people. This is probably a mistake. If other functions take up a disproportionate amount of time, ordering is usually the first to suffer. It is preferable, therefore, to allocate the function to a specific clerk or set of clerks.

The task of ordering is greatly facilitated if a mark is placed on each card when stock falls to preset levels. The mark should be placed accordingly by the person recording an outward movement. This is therefore the job of the clerk responsible for entry of details from customer invoices. Order clerks then pick up low-stock information from marked cards and remove the marks after re-orders have been placed.

Parts should be supplied in a regular flow which may vary with seasonal demand. Regular, frequent ordering cuts down the size of stock investment needed, reduces the required storage space required and evens out the workload.

In a monthly ordering cycle, order clerks review the stock cards once a month for items that have reached the order point. This is relatively simple if these have been marked by the entry clerks. A refinement of the system is to pick a specific day in the month for each section of stock.

A weekly ordering cycle is the most highly disciplined. This system utilizes a manufacturer's order reception system which is open to receive telexed orders from a certain list of distributors only at a specified time each week. If that time is missed, the order must wait until the following week. Naturally, only a small number of items will be needed each week, so the task of compiling the order can be completed rapidly. Under this system the flow of parts is rapid with very few in stock and many in the pipeline. The detailed information needed to make up the order may be assembled on a form designed for this purpose. This form, the "Stock Replacement Indent Form" and its uses are discussed further in Chapter 3.

Once a stock order has been compiled it is typed, either on the manufacturer's special order forms or on letterhead or forms of the organization placing the order. Some manufacturers insist that their own forms are used since this speeds up documentation in their administrative processes. Normally, goods are dispatched as soon as possible on credit terms mutually agreed between the two parties. If normal credit cannot be provided, an alternative payment mode will be required.

2.4 Payment terms

There can be several reasons for lack of credit. A manufacturer may need cash and be unable to advance credit, or the enterprise placing the order may not be considered creditworthy. If the order comes from a foreign country, there may be factors outside the control of both parties, such as exchange control or import licensing regulations, which make normal credit impossible.

If the problem is one of cash flow for either the purchaser or the supplier, assistance may be obtained from financing institutions. A bank may lend cash to the purchaser or it may pay the supplier directly in exchange for the invoices, which will subsequently be used to obtain payment from the purchaser.

Other financing agencies, such as confirming houses, combine the functions of credit provision with services such as ordering and eventual supervision of shipment of the goods. The confirming house is a credit-worthy institution, usually located in the manufacturer's home country, which buys on normal credit terms from the manufacturer. It extends a line of credit to the purchaser based on credit worthiness and on the credit terms advised for the purchaser's country by the official export credit guarantee service of the country in which the confirming house is based. Covered by the export credit guarantee service, the confirming house is then able to sell its invoices to a bank to finance the operation.

The use of a confirming house is attractive in that all or most purchases may be made on advantageous terms (often up to six months). However, many suppliers are involved and orders from each individual supplier may not be sufficient to warrant such credit. Also, interest rates in the country of supply (plus the confirming house commission of up to about 3%) may be much lower than those in the country of purchase.

Import licence and foreign exchange regulations of the purchaser's country and occasionally that of the seller may obstruct the regular flow of orders and payments. Import licences are often a means of limiting foreign exchange outflow as well as protecting local manufacturers. If an import licence is required, the conditions under which it is granted must be ascertained; there may be no alternative to buying certain items from a domestic manufacturer. On the other hand, if the quality of local parts is below the required standard, a case can often be made for a licence.

When a license is needed, parts should be selected carefully for maximum advantage of the limited foreign exchange quota. Often, such quotas are based on the volume of previous business, so the enterprise that has a good parts support history will be in a better position. For import licences and other foreign exchange regulations, it is usually necessary to know the value of an order before it is placed. In this case the parts list must be sent to the manufacturer for pricing and will normally be returned as a "proforma invoice", an invoice including all ancillary expenses such as packing, freight and insurance, presented as a quotation in advance of any formal supply contract or order.

The proforma invoice may be needed to obtain an import licence or for advance booking of foreign exchange. It may be part of a control system in which goods are subject to independent inspection and value analysis before dispatch. The proforma invoice is also necessary if payment is a letter of credit.

In international commerce, there are several ways in which suppliers are paid for goods shipped overseas. They vary from the confident assumption that the purchaser will always pay to the most cautious suspicion that the purchaser will pay only if the most stringent conditions are applied. The degree of stringency depends first on external factors, such as the ability of the purchaser's country to meet its foreign exchange obligations and second, on the relationship and degree of trust between purchaser and supplier.

The easiest payment system involves goods sent "on open account". This implies that the supplier expects payment within a period of 30-60 days from the date of shipment, depending on the time it takes for the goods and the documents to arrive.

Expediency of payment is largely a moral decision but also depends on the purchaser's desire to maintain favourable credit terms. Another consideration is the supplier's ability to take legal action to obtain payment if it is not made on time. Under some circumstances, open account terms may be extended to a period of 60 or 90 days with interest either added or waived. If no interest is applied to credit terms however, it may be assumed that it is absorbed in the price of the goods.

Under the "cash against documents" (CAD) system, payment security is obtained on condition that payment is received in exchange for the documents giving the purchaser title to the goods. The "documents" are invoices and transport papers, plus a bill of exchange. Transport papers are bills of lading in the case of ship's cargo, air waybills in the case of goods sent by air, or a road or rail transport consignment note. In any case, payment must be made prior to receipt of the transport papers. When the purchaser receives title to the goods, they become his property. For this reason, it is normal for documents to be sent through a bank.

In some cases the supplier may immediately receive cash from its own bank for the value of the invoices (less a discount). This bank then sends the documents to a correspondent bank in the purchaser's country. The purchaser can obtain the documents (and thus possession of the goods) only by making payment to that bank. In a variation on this system, a period of credit may be allowed during which the purchaser "accepts" the bill of exchange for payment on a specified date. If payment is not received on that date, legal action may be taken locally by the bank to recover it on behalf of the supplier. Banks charge fees for this service based on the amount of time and effort involved. Where legal action is taken, charges can be high and the supplier can be without the funds for a prolonged period. Even when the terms are strictly "cash against documents" and the purchaser does not have the necessary funds, he may simple leave the documents with the presenting bank. In this case, the goods are left in the hands of the transport company or are obtained by other means. The system of "cash against documents" or "acceptance of documents" is thus not entirely secure from the supplier's point of view.

The method suppliers regard as most secure (except for cash in advance, which is very rare) is the "documentary credit" or "letter of credit". In this case, the purchaser makes an advance commitment to a local bank which then pays a correspondent bank in the supplier's country. The correspondent bank pays the supplier or offers to collect on behalf of the supplier. In the first case, the letter of credit is said to be "confirmed" for payment by the supplier's bank. In the second case, it is "unconfirmed" and payment depends not only on the ability of the purchaser's bank to pay (which is seldom in doubt) but also on its ability to obtain the necessary foreign exchange, a factor which can result in delays from a few weeks to several years.

The decision to confirm a letter of credit is entirely in the hands of the correspondent bank in the supplier's country. There is nothing that the purchaser or its local bank can do to influence that decision. Suppliers can chose to rely on payment through a confirmed documentary credit, but if it is not confirmed they may regard it as no more secure than any other form of credit.

Since a letter of credit is a commitment by a bank to pay money, it contains payment and documentation conditions and is normally only valid if all of these conditions are met within a specified period of time. The banks concerned have no control over whether or not the supplier can meet all these conditions, and for this reason do not regard letters of credit as collateral security on which they can finance a supplier. The purchaser can help by insisting that its own bank apply a minimum of conditions to the credit and a reasonable time limit for shipment. Partial shipments should always be allowed and unless there are any real obstacles, transhipment should also be allowed. Letters of credit and other documents controlling shipping and payment are generally covered by rules and regulations established and published by the International Chamber of Commerce.

2.5 Customs clearance

Once purchasers have title to the goods, they have, in effect, taken possession of them. In the case of an importer, however, the goods must be cleared through customs at the port of entry. This can be a tedious process because of the detail involved in analyzing the various duty rates on different categories of parts. This may apply even with agricultural machinery, even though most agricultural machinery parts are classified duty free, as customs officers tend to be skeptical about the ultimate destiny of parts which could also be used on high-duty motor vehicles.

A large organization may deal most efficiently with customs clearance by employing clerks specializing in this function. Good candidates for these positions are former customs officers who possess the detailed knowledge of customs procedures and paperwork that are crucial to rapid clearance of the goods.

Port and especially airport operators allow only a very limited period for goods to remain on their premises. If goods are not cleared quickly, demurage charges may have to be paid. There may also be port charges for heavy lifting or other handling costs, customs fees and stamp duties for processing documents and overtime charges for customs officers, especially where consignments are very large. All of the incidental charges for a parts consignment should be recorded on a specially-designed form and kept in the order file. These costs should be summarized periodically and calculated as a percentage of the invoiced value of the parts. Not only is this a valuable part of the pricing process, it provides material for cost reduction studies. It is essential to know the real cost of purchasing and incidental charges can add up to a significant proportion of the total.

The establishment of an internal customs department is not justifiable for a small organization. The break-even point for this decision is where the task of clearing incoming goods provides a full-time job for one person. Below this point, it is probably more economical to employ a clearing agent - a freight forwarding company that specializes in the movement of goods.

Care should be taken in choosing an agent. Freight-forwarding companies are often started by retired customs officers and it is wise to check that retirement was not forced by incompetence or fraud. The service of international freight forwarding companies may be less personal; tasks may be given to junior clerks without adequate supervision. However, if the

Forwarder's Air Waybill

Forwarder's name Address			House Airbill No.		
Airport of Departure		Routing	Destination Airport		
Consignee's Name & Address		Also Notify			
		Flight	Day		Service
Shipper's Name & Address		Multiple Air Waybill No.			
Shipper's Order No.	Account No.	Customs Reference No.			
CERTIFICATION: WE HEARBY DECLARE that the goods mentioned herein were dispatched by .... on .... by flight .... Carrier certifies goods were received for carriage subject to the terms and conditions of the Institute of Freight Forwarders Ltd. Signed............as agent for issuing carrier.					
Number of Packages	Actual Gross Weight Kilos	Nature and Descript.	Chargeable Weight	Rate	Total
Dimension (cm)	Marks and Numbers		Letter of Credit Details Other Information		
Currency	Shipper's Declared Value for		Docs. to Accompany Shpt.		
	Carriage	Customs	Insurance	Consular Inv.	Commer. Inv.	Certif. of Orig
Invoice Details Freight (see above) Inbound Freight Insurance Local Carriage Local Processing Disbursements C.O.D. TOTAL		Prepaid	xxxxx xxxxx	Collect	xxxxx

importer is familiar with the procedures and is prompt in paying forwarder's bills, then efficient service may be expected.

Clearance procedures cannot start until the importer is in possession of an invoice, but they can start before title to the goods has been obtained. In fact, customs pre-clearance has become the rule where ports (especially airports) have become congested. In this case, the importer, armed with the supplier's invoice, clears customs formalities and pays whatever duty is required before the goods are shipped. Once cleared, the goods are shipped and the importer may remove them as soon as title is received.

Early clearance implies that the importer must have at least one invoice to begin preparation of customs clearance papers. In the case of pre-clearance, this is no problem. Where goods are invoiced and shipped normally, however, there may be delays in arrival of the documents. Where journeys are short, the goods may arrive before the documents. To overcome this problem, arrangements can be made for the documents to be shipped with the goods or by courier. Banks may or may not agree to do this. If not, a separate invoice may be sent by courier which will not give title to the goods but will enable clearance procedures to begin.

2.6 Emergency orders

Emergency orders are subject to all of the above provisions. They too become a legal contract between buyer and seller with all the usual conditions regarding terms of payment and documentation and they must be cleared through customs before the goods can be made available. Reputable machinery manufacturers have special procedures to accelerate the processing of emergency orders. A good system dispatches emergency orders received by mid-day on the same day, and those received after that on the following day. Some suppliers guarantee dispatch within 48 hours rather than the same day. A price premium or reduced discount may be applied to emergency orders or certain categories of fast-moving which the manufacturer believes the local stockist should not run out of.

For the emergency order to work, prior agreement must be reached on payment, and import licenses must be available where required. Many manufacturers accept a relatively low-volume of emergency orders (the average is about five lines) on open account terms. Where this is not feasible, the use of a confirming house greatly assists the process because suppliers will ship in response to a telephone call or a telex from a confirming house without waiting for formal orders.

When the parts arrive at their destination, clearing them quickly primarily depends on the combined efforts of the supplier and transport agency in getting the documents to their destination at the same time as the parts - often in a special pouch.

2.7 Back orders

A supplier normally provides a service level in excess of 85 percent - some go as high as 97 percent - which means that 3-15 percent of the parts ordered will not be shipped in the first consignment. Events such as strikes may reduce a supplier's stock below normal levels, or parts may be shipped in reduced quantities. The result is a back order. Generally, the missing parts are shipped within a few days, but it can take weeks before they are available. In the meantime, import licences and letters of credit - even those allowing partial shipments - can expire.

There is therefore a need for close liaison between supplier and purchaser on back orders. Some suppliers cancel outstanding items, either immediately or after a specified period, and ask for a re-order. This is a good system, but must be carefully monitored. Otherwise stock cards will show an exaggerated quantity "on order" which will distort future orders. Also, where a letter of credit is involved, the purchaser may have to deposit the full value in local currency. If the full value of the credit will not be utilized because some parts are on cancelled back order, that portion of the deposit must be recovered, not always an easy task.

Interpreting the flow of invoices representing either major parts of orders, back-ordered portions or both is not difficult once the supplier's invoice pattern has become familiar. This is an important part of the clerical work of the parts department. All of the information must be transferred to the stock cards - a task which may be performed by the same clerk(s) who compiled the orders, or by a separate section devoted to recording incoming goods. In any case, the invoices should not be entered on the cards until the goods are physically checked into the store.

The supplier's invoice may vary in appearance. Sometimes a commercial invoice form is used for all customers, plus any special invoices required by the customs regulations in the country of destination. There has recently been some standardization in such documents, but it is by no means complete. Often a supplier will use the special or standardized customs invoice as a front page to show the general category of parts and freight and ancillary shipping charges. Details of the parts and their prices are given on commercial invoices on subsequent pages. Examples are shown in Appendix A.

2.8 Stock administration

When parts arrive, they should be accompanied by a packing list and a copy of the invoice should be made available to the parts department for checking purposes. The warehouse staff should unpack, identify and physically check all incoming parts. Parts may have been omitted by the supplier or even stolen. In the case of missing parts, a short shipment claim must be sent to the supplier. There is usually a time limit for submission of such claims; if the supplier is not notified immediately, it may be suspected that parts have been mislaid at destination.

In case of theft or damage, an insurance claim may recover the value of the parts. Again, such a claim

Parts Invoice

```
Name (Of organization invoicing the parts and using
      pre-printed form sets of which both the order
      and the invoice are parts).

Address


To: (name of buyer)          Order No. _____
                             Date _____
                             Dispatch by _____
Address:                     Mark:


Send to:



Terms of Payment:            Invoice No. _____
Account No. _____       Date _____

Dispatched by:               Date _____
Carrier:                     Way Bill No. _____
```

Item	Description	Part No.	Qty.	Price	Total	Bo

```
                             Total Parts
Signed:                    + Carriage
Date:                      - Discount
                             TOTAL

          Page ____ of ____
```

should be made immediately and all evidence, e.g. damaged packings, retained for inspection. The passage of time erodes the credibility of such a claim.

The location of parts will depend on a number of factors such as size, weight, rate of movement and relationship to other parts. In some stores, all parts relating to one supplier may be kept together, whether or not the location is convenient for picking. In other stores, parts are kept in strict numerical sequence, whether or not this separates closely-related parts. Regardless of the type of system used, the location of every part in the store must be recorded on the inventory card in the card index.

The best location plan, and the easiest one to administer, takes into consideration the fact that at least 80 percent of movement comes from less than 20 percent of the parts. This portion of the stock, the common, fast-moving items, should therefore be located where it is most accessible. It is then natural to group related parts - nuts and bolts, pistons and piston rings, etc., as their usage rates are similar.

When a store is first set up, it may take some time to sort out just where parts are best located and which parts are fast movers. There may be some shuffling between bins and shelf units. This may be minimized by leaving plenty of space on each shelf so that re-grouping will not require a new location for everything else on the shelf. Every move means that the location reference on the card must be changed, so it is advisable to restrict moves to a minimum. However, a massive re-arrangement is sometimes necessary to phase out old models and introduce new ones. In this case, it is advisable to use a location change record sheet.

All new locations should be recorded on the inventory cards and the record sheets should be kept for future reference in case of card entry errors.

The bin location numbering system should be simple but comprehensive. If a system of letters and numbers is used then the shelf units may be labelled A,B,C, etc.; the shelves 1,2,3,etc.; and the positions on the shelves given a second pair of digits. If a part is located at position A6.16 it will be found in unit A on shelf number 6 at shelf position 16. For this purpose, the shelves should be divided into sections of fixed length, i.e. 10cm, marked with a spot of paint on the edge of the shelf. Each section will be numbered. It does not matter if a part occupies more than one section.

Small parts stocked in small quantities should be kept in boxes commonly 10 cm wide. Several different parts may be kept in one box if they are individually packaged and clearly identified.

Shelf unit sections are generally 1 m long. Where rows are long, it may be advisable to identify the sections separately. Row A may thus have sections AA, AB, AC, etc.

Special storage units such as racks and pallets can also be identified by row letter and a set of numbers indicating vertical and horizontal position in the row. The value of this system is that every storage position will have a set of co-ordinates which clearly identifies its logical relationship to its neighbours.

Re-arrangement of parts will not create a need to re-number shelf spaces. Re-arrangement of shelving to obtain more storage capacity will require some re-numbering unless all the units and shelves are simply moved up and new ones inserted after the last one. Such moves often accompany major re-organization of parts and any new locations can be recorded and transferred to the cards using a location change record sheet.

Normally, the inventory card will be the only parts location record. Under some circumstances however, it is advisable to use bin cards, inventory cards kept with the parts in the bin or on the shelf to record movements, with dates and references. The use of bin cards has largely been phased out by the development of good inventory control systems based on a central card index or computer. However, they still can be valuable in a subsidiary store when the inventory control card index is kept in the main store some distance away. Inventory control functions are still carried out on the main card index. The bin cards serve only as a check on how much stock should be in the bin, and as an additional source of reference if records are lost in transit between the main store and the subsidiary.

2.9 Stock checks

Store administration normally includes an annual "stock taking" to satisfy management and auditors that the value represented by the parts stock in the accounts is actually present on the shelves. There are two methods of stock-taking. The standard system is to close the store, or a section of it, for as long as it takes to make a physical count of all parts. The physical count is then compared with the balance shown on the cards. If there are only minor differences, the cards can be changed to reflect the actual quantity. If, on the other hand, there are major differences - especially deficits - then an investigation is necessary , and all references must be checked to isolate the error. If, after full investigation, the missing parts cannot be traced, they must be written off the accounts as lost value. Depending on the frequency of such deficits and the operating systems used, management may take action to improve clerical performance, or even replace staff.

Stock quantities above the theoretical level represent error just as much as deficits and unaccounted stock is not a bonus - the auditors do not count it that way even if it balances the deficits. It will usually have been paid for even if it is not on the cards.

The store closure for stock-taking represents a probable loss of business and if the job is done after hours, it is expensive in overtime payments. After the physical count, the investigation of differences can take weeks and the accounts cannot be closed until it is complete. Annual stock-taking often represents a severe disturbance to the flow of activity.

Continuous stock-taking throughout the year avoids the trauma of the annual exercise. Before a decision is taken to change to continuous stock-taking, the advice of the organization's auditors should be sought since they may have conditions to apply. For example, they may require that stock-taking be done by someone other than a member of the regular departmental staff. This system is known as perpetual inventory.

If the organization is large enough, it may require a team to maintain the continuous check. The team leader should have some accounting experience and be used to tracing documentary references. A clerk is needed to write down the data. Manual labourers may be needed to handle the parts.

The perpetual inventory method has several advantages. First, fast-moving parts and valuable items can be checked frequently; the slow-moving and less valuable stock perhaps only once a year. In addition, sections of stock can be chosen at the last moment so that the regular staff do not know which section is to be checked next. This can help security. Apart from the official stock-taking, it should be the parts manager's duty to make spot checks in the same way that other sections of the enterprise are checked in the course of daily supervision.

It is the warehouse supervisor's job to keep the store in good order. Parts should be on the shelves where they belong, arranged neatly with the part num-

bers visible. The floor, especially the aisles between shelving units, should be clear and clean. Empty cartons or loose packing materials should not be left on the floor. They impede movement and are a fire hazard.

When manufacturers ship parts in protective packaging, it should be left intact if possible and the part number written on the packet. Opening the boxes leaves the parts vulnerable to dust and humidity – hazards in tropical climates.

Neatness and cleanliness are necessary for efficient operation and the maintenance of parts in good condition. An industrial vacuum cleaner is useful for shelves as well as floors.

2.10 Removal of parts

Removal of parts from the warehouse should be stringently controlled and training programmes for warehouse staff should feature such controls.

The process begins with the receipt of a stock-picking document in the warehouse. In some computerized systems, a picking slip is printed for each part needed. In less sophisticated systems, a copy of the customer's order may be used. In the system recommended in this publication, the customer's order is part of the set of documents that eventually produce the invoice. The warehouse supervisor gives the document to a warehouse worker who picks the parts from the shelves and brings them to the dispatch area.

In a large store, warehouse staff may be divided into two teams, one to deal with incoming goods and one to pick and dispatch stock. The packers for dispatch may even be separated into a third team. Equipment such as forklifts may be available to all or allocated individually to each team. In a major supplier's warehouse, the staff may even be further sub-divided into teams dealing with small as distinct from large parts; teams which pack small cartons as

Parts Picking Document

Name (Of organization invoicing the parts and using pre-printed form sets of which both the order and the invoice are parts).						
Address							
To: (name of buyer) Address:	Order No. _____ Date _____ Dispatch by _____ Mark:					
Send to:						
Terms of Payment: Account No. _____	Invoice No. _____ Date _____					
Letter of Credit No. _____ Import Licence No. _____	Expiry Date _____ Expiry Date _____					
Item	Description	Part No.	Qty.	Picked	Locn.
Signed: Date: Page ____ of ____						

distinct from those who crate large consignments. Emergency orders are usually handled by a special team in such organizations.

In a small warehouse, these functions may be combined but the distinction between procedures for incoming goods and those for outgoing goods should be maintained. The floor areas in which they are performed should be kept separate to avoid confusion of partially packed and unpacked, and incoming and outgoing consignments.

When the parts have been picked and the assembled consignment has been checked, the picking document is marked to show the quantities of parts supplied and those "to follow", if any. The parts, accompanied by the marked-up picking document are then passed to the counter for invoicing and release to the customer. There is no need for clerks in the warehouse itself as the only clerical job performed in the warehouse should be the final entry of quantities on the picking document.

2.11 Sales administration

Selling parts is not necessarily simply stocking and responding to an external demand. In a commercial organization, an active marketing policy for parts can boost profits. In a project or government department, although there is generally no need to sell parts actively, the parts manager can sometimes persuade the service manager to expand the fleet overhaul programme in order to clear out some slow-moving stock. The principal function of parts sales administration is to ensure that the detailed requirements of the machinery user are correctly interpreted, thus maximizing customer satisfaction.

In any organization, the attitude and performance of the staff dealing with customers makes a great deal of difference in the department's overall effectiveness.

A counter where customers present their orders should be set up in an area separated from the counter used by the company's own workshop staff since different procedures are used in the two situations. In addition, a more comfortable atmosphere conducive to selling can be maintained in an area not congested with mechanics in dirty overalls. The counter is also an important interface between the parts department and its customers and is not only the point where most transactions take place, it is also the point at which the identity of the parts needed is defined.

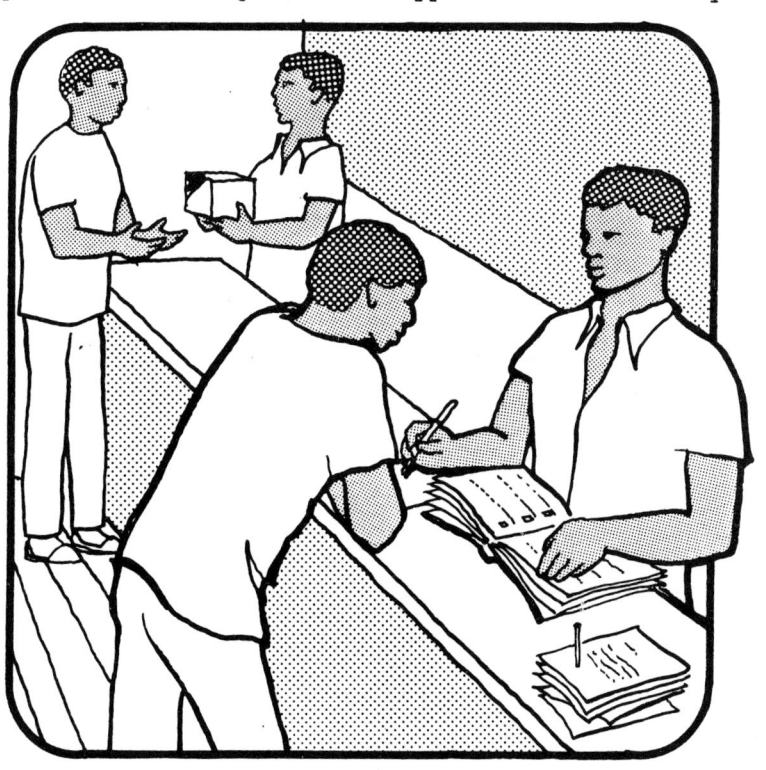

In many cases, customers - internal or external - will not have the complete identity of the part they need. They may or may not know its proper name. There may be several terms used commonly for the same part. For example, some types of oil seal may be described as "sealing ring", "washer", "gasket", "oil seal" or "oil ring". When a part is sold in a foreign country, the problem is compounded by an equal or greater variety of expressions in the second language. Further difficulties may be created by the fact that many parts look alike. Sometimes the differences in specification and fit are not even visible.

Correct parts identification at the counter, therefore, depends on a high standard of accuracy in interpreting the customer's needs from the parts manufacturer's catalogue. The counter should be provided with up-to-date catalogues in the form of books or microfiche sets, and the equipment for reading them. This will give the counter staff all the information necessary to identify parts required for a specific machine using the manufacturer's terminology and its vital part number.

Up-to-date knowledge of the machinery involved is required for parts identification. Most manufacturers maintain continuous programmes of cost reduction or product improvement, or both. Since this involves changing parts, either individually or in groups, the effects on the task of parts identification are ongoing. A new machine model may require much modification before it becomes relatively reliable. During the first phase, parts changes may be massive. The second phase may bring less frequent changes. In the third phase, when the model is obsolete, there may still be some rationalization, and sub-assemblies which go on future models may still be subject to change. Before the microfiche, parts changes were usually publicized through parts bulletins or notices, and parts catalogues often had to be up-dated manually. In an organization using parts books at several work stations, this could produce employment for a whole clerical section. Periodically, a new version of the parts book would be issued by the manufacturer and the process would go on again.

Microfiche is cheap enough to print new fiche from an up-to-date master. Instead of amendments to printed books, new fiche are sent out on a world-wide mailing list. There are several standards used for microfiche requiring different degrees of magnification. A parts department serving machines from several manufacturers may need several microfiche readers.

In addition to helping the customer identify the exact part he needs, the counter clerk also can suggest other parts which may be needed in association with the part requested. For example, it would be foolish to buy a new crankshaft without also purchasing new bearings and a set of relatively inexpensive parts, such as oil seals and gaskets, which would enable the crankshaft replacement to be completed with less risk of recurrent problems. Also, some parts have to be replaced in matched sets; for example, a new differential crown wheel almost always needs a new matching drive pinion.

This type of knowledge is vital to the parts counter staff if the long-term interests of the machine user are to be served. It is thus a highly skilled job, and one where careful selection and continued training are required. The counter staff need diplomacy to deal with customers angered by shortages, by waiting or by prices about which they often have unrealistic ideas. Sales techniques are required in a commercial organization and reasonable skills in arithmetic are needed when pricing is carried out at the counter.

All the customer's requirements should be written onto a customer order form which may be part of a set of documents including the final invoice and the picking document. The next step is to search the inventory and index for information. Where the parts are available as requested or can be substituted by updated versions, the quantity required is entered on the order, (and hence on the picking slip and the invoice), together with the unit price. The picking document is then passed to the warehouse staff.

If parts are not available, decisions have to be taken between the counter staff and the customer. It may be possible to find substitutes or to switch between individual parts and complete assemblies. The local market may be an alternative source. If it is necessary to raise an emergency order and bring the part in specially, prior agreement is essential. Otherwise, when the part arrives with bills for emergency service and air freight, it may be found that the customer has made other arrangements.

Back-orders have to be kept under continuous review and communication is just as important with customers as it is with suppliers. Incomplete customer orders should be kept in an open file. The review procedures and frequency should be clearly laid down. They may be delegated to the counter staff or be specific responsibilities of the sales supervisor, or even the parts manager. Availability and pricing information are drawn from the inventory record cards in traditional systems. In a small store, the counter staff may have direct access to the cards. In a large store, they may have to pass the customer orders to a special clerical section set up to deal with them.

The use of computers in stores administration is spreading rapidly. Counter staff may have visual display units (VDU) from which they can call up all the information they need by entering the part number on a keyboard. The system may allow the picking slips to be printed along with an invoice. The customer's account may be debited with the resulting invoice total. In less sophisticated systems, the inventory is tabulated by the computer and printed out as a counter catalogue. Customer orders can be priced from it, but stock availability will still have to be checked and the invoice calculated manually.

It is quite common for counter staff to handle cash - usually through a till. In some cases, however, counter staff are selected with an education level which does not adequately prepare them for this responsibility. Under these circumstances, a cashier may be appointed to make the final invoice calculations and to receive and account for all cash. This system also has security advantages since the counter staff can be given open access to the public while the cashier is provided with a secure cage with limited public access, and a strong box or safe for the money.

Not all sales are made over the counter. A commercial organization will have customers who order by mail or telephone. Government departments or projects may have outlying stations which depend on written or telephoned communications.

Mail orders may be handled by the counter staff before or after, or in between, dealing with people at the counter. In a large organization, a special section may be established to deal with mailed orders. Telephone orders are usually handled by counter staff and here the customer order form provides a framework for efficiency. Otherwise there is a tendency to scribble notes and important information is frequently omitted.

Where major customers do most of their purchasing by mail, and the only personal contact with the parts department is via the driver who picks up the goods, there is a strong case for regular visits to the customer by the parts manager or one of the sales staff. In this way, new orders can be obtained, back orders can be reviewed, problems discussed and outstanding accounts collected. Building goodwill need not be limited to commercial operations. Even in a government department or project, efficiency can be improved by contact between headquarters staff and outlying stations.

2.12 Clerical functions

The clerical staff form the third group of people in the parts department organization. They deal with suppliers and customers, and keep records of various kinds. There may be several sub-groupings.

In a small store, the function of the order clerks may include responsibility for the stock record card index. This index is the central memory unit and holds all the vital information about parts in the store and which parts will need to be ordered to keep customers satisfied. Good administration of the card index is therefore vital.

Card index staff should be selected for good attention to detail, accuracy, numeracy, and reliability. They will need close supervision and the clerical supervisor should spend part of each day spot checking the cards. Otherwise, the management reports will show a drop in customer service, and at stocktaking there will be a massive task of sorting out deficits and unaccounted excess parts.

Information must be entered accurately on the right cards in the right order. The more it can be subdivided into specialist functions, the better. In a large store this means separate teams entering incoming goods, outgoing goods, orders and the updating of prices or technical information.

Even in a small store, it should be possible to systematize the various functions. For example, the first task every morning could be the entry of all the previous day's issues. In the middle of the day, new pricing information, part number changes, and changes in bin location could be entered. The last task every evening could be the entry of all goods received during the day. Specific times or specific days can be scheduled for ordering.

A separate section of this chapter deals with the techniques of inventory management. It concerns the task of ensuring that information is displayed at all times on the cards. The entry of issues and receipts must be completed daily. Unless the stock balances on the cards are kept up to date, the warehouse staff may waste time searching empty bins and customers will take a sceptical view of an organization that seems not to know what it has in stock. If the cards are not kept up to date, re-ordering cannot be done accurately. For example, an order will not be placed if the card shows a balance greater than the minimum. But the balance will not be accurate if sales invoices for a part are awaiting entry on the cards, while in the warehouse the bin for this part is empty.

The administration of the record card index involves the recording and calculation of a mass of small details. Details of parts received are taken from the manufacturers' invoices by the warehouse staff responsible for receiving the parts. Details of parts issues come from customer invoices and workshop requisitions. These also give details of backorders which show up on the cards as negative balances. Information on orders placed for parts will be taken from copies of the orders after they have been approved for dispatch to the suppliers.

New part numbers, prices, minimum stock, order point and similar entries, come from a variety of sources. No information should be entered without documentary confirmation, whether in the form of a manufacturer's notice, or a memorandum from the parts manager regarding changes in re-order procedures.

To ensure that all documents reach the card index clerks, there should be a formal system of document routing which begins with the creation of the document and ends with its storage in a filing cabinet. The chart at the beginning of this chapter shows the paths taken by orders, parts invoices and payments between the parts department, its suppliers on the one hand and its customers on the other.

The organization's size and complexity determines some aspects of document routing. For example, in a small organization departments will be fairly close together and it may then be unnecessary for the parts department to keep file copies of invoices for its own use since they will be lodged in the accounts department and may be referred to there. In a larger organization, copies of invoices should be kept in the parts department, particularly invoices which may be needed for research or problem-solving.

Document routing can be illustrated by flow diagrams. Variations may be designed for individual organizations, and a written procedural manual may be used for training purposes. It will also be needed if key staff members are absent and others have to take over. It is also a starting point for any overhaul of the management system. The steps that have caused problems or inefficiency can be identified and corrected. A new version of the manual can then be published.

2.13 Inventory management techniques

Stocking is limited by space and cost. A project or a government department has a limited budget for parts stock. A commercial enterprise is limited by the selection of parts which can be sold relatively rapidly and profitably.

The parts manager is responsible for deciding what to keep in stock. In the case of a new set-up, the parts manager may find that decisions on initial stock composition have been taken by government administrators or company executives. These decisions are based on several factors including ease of future re-ordering, available funds, and the type of machinery to be serviced.

In other cases, the first task of the parts manager may be to decide on the initial stock himself. In this case, the parts manager will need to use all of his skills, experience and financial management ability to set out rules and guidelines to provide staff with a framework within which orders can be formulated.

A large element of the skill in parts management is the ability to forecast correctly which parts will be required in the future. A low estimate saves on stock investment, but eventually costs more in emergency orders. A high estimate costs more in stock without necessarily paying off in sales. The best clue to future parts demand is generally found to be recent parts demand, but inevitably this reflects unseen variables which may not be repeated.

Some variables can be ironed out by comparing past demand over a period of years. The more the figures can be broken down, the more the variances can be isolated. If sales of a part per day throughout each of the previous five years were to be studied statistically, a high order of accuracy might be achieved, but for each part this would mean analysis of more than 1 500 records. Calculation from weekly records is still too costly for the degree of accuracy required, except in circumstances where movement is very rapid, and weekly orders are placed.

For most purposes, analysis of monthly sales figures gives a sufficiently accurate forecast and is convenient for a monthly ordering cycle. Strictly speaking, a minimum of 30 records is required for valid statistical analysis, but unless the variations are very large, records of sales over two or three months will give at least an idea of the rate of movement. Large variations are easily identifiable and often also easily explained. For example, in the case of parts for agricultural machinery, the seasons cause predictable variations. Certain types of parts are needed during the active equipment operation season - or just before it if users anticipate their needs. Other types are needed for overhauls outside the busy season. Seasonal peaks in the parts for the transport industry may be caused by large amounts of agricultural produce to carry at harvest time, or by varying road conditions.

Another measurable factor which affects parts sales is the age of the machines which are being served. Machines require a relatively small investment in replacement parts during their first two to three years of operation. As the machines age and parts start to wear out, the need for parts accelerates. Finally, as the machine's value is reduced by depreciation, the cost of maintenance increases to an uneconomical level, and the need for parts diminishes as the machine goes into semi-retirement or is sold. An experienced parts manager who knows the state of the local machinery population will be able to judge how stocks should be modified.

As a rule, experience has shown that 80 percent of movement comes from only 20 percent of stock. This is more or less true whether movement is considered in terms of numbers of parts or value. Conversely, 80 percent of the stock provides only 20 percent of the movement. This portion cannot, however, be ignored because the lack of one part in this category can prevent completion of a repair job that uses mostly fast-moving parts. This phenomenon is known as Pareto's Law.

Pareto's Law has been used to develop a management guide for a system of parts organization known as the ABC/XYZ system. Parts are categorized A, B or C according to their rate of movement, measured as sales value. The 20 percent of the parts stock which represent 80 percent of the sales volume and value - the most important parts - are classed as category A. Category B parts account for about 5-10 percent of the total in terms of numbers and 15 percent in sales value. The remaining parts - those responsible for only 5 percent of sales value - are classed as Category C.

The stock is then re-classified according to cost. The items making up 80 percent of the cost are called category X. In the best of cases, these will also be category A parts from a sales point of view; in any case, they are parts to be kept under strict control. Category Y parts make up the next 15 percent of stock value and are also important, while category Z parts, however numerous in type and quantity, only account for 5 percent of stock value and do not merit much attention.

The value of this system is that it enables management to concentrate on the essentials without becoming bogged down with unimportant detail. Also, if the analysis shows that many parts are in category CX, there is obviously an imbalance in stock with too high an investment in slow-moving items. Both categories should be shown on the cards and rules can be laid down about how they are to be treated.

As soon as an organization starts operating and management gains some experience of the parts that move and those that do not, a simple chart which categorizes parts according to their value and rate of movement can be used to lay down the ground rules for ordering.

It is normal to consider a part slow-moving if three units or less are issued per year. Between three and ten units, a part is a medium rate mover. More than ten units a year classifies a part as a fast mover. Low-value parts are those which cost less than $5. Medium-value parts cost between $5 and $150. Anything costing more than $150 is a high-value part and should not be stocked at all if it is also slow moving. It will be cheaper to bring in this type of part only as needed even considering the

Parts Ordering Guide

		Movement		
		Fast	Medium	Slow
V A L U E	Low	6 mo.	9 mo.	12 mo.
	Medium	3 mo.	3 mo.	1 unit
	High	3 mo.	2 units	none

additional cost of an emergency order. These values and movement rate classifications can be varied to suit the type of part and other local factors. The objective is to stock economical, fast-moving parts while restricting the purchase of expensive items that are required only on rare occasions.

The chart indicates that low-value, slow-moving parts should be purchased in bulk quantities representing a whole year's sales. This is because the preparation, dispatch and other handling costs of an order, can amount to more than the cost of the actual parts if they are bought in small quantities. On the other hand, available space must be considered as well; a year's supply of inexpensive but bulky items, might cost so much in space "rental" that the economy of bulk purchase is reversed. The true cost of holding stock depends on many factors, including interest rates, insurance, obsolescence, maintenance and overhead costs.

For small stores, the simple guide chart will be sufficient for judging how much stock of a part to keep or to order. If the store has a computerized inventory control system, however, it is relatively easy to build an equation into the programme which calculates the "Economical Order Quantity" (EOQ) for each part each time it is ordered.

This calculation, takes into consideration three factors: annual sales, cost of holding stock and cost of placing an order. Stock-holding costs and the ordering cost may be treated as constants (which may be changed when factors such as interest rates change). The single variable which the programme will use to calculate the EOQ will then be the estimated annual sales figure for the next 12 months.

As the quantity ordered at any one time increases in relation to annual sales, the average quantity held in stock will also increase and the cost, per item, of holding it in stock will rise in proportion. Conversely, as the quantity per order increases, the cost of ordering decreases. If these two calculations are illustrated graphically, it can be seen that there is a minimum cost point where the two lines cross; and a third curve can be plotted to show the total cost.

The average stock level (ignoring safety stocks), will be half the order quantity. This is based on the assumption that stock is zero just before a supply arrives, and falls to zero again immediately before the next order arrives; over the period between orders, therefore, stock value will average half the order quantity value.

If the order quantity is Q, the average stock will be Q/2 and the cost of holding this stock will be \$cQ/2 where c is the cost factor (expressed as a percentage of the total stock value). The cost of placing an order, represented by \$x, must be multiplied by the number of orders placed which is the annual quantity sold, (S), divided by the order quantity, (Q), or \$xS/Q. If the most economical order represents both the minimum order cost and the minimum stock-holding cost, it must be at the point where \$xS/Q = \$cQ/2 which is the point where the two lines cross on the graph. The order quantity Q is therefore equal to the square root of 2xS/c.

As an example, for an item where annual sales are \$1 000, ordering cost is \$0.20 per order and stock holding cost is 25 percent per year, the effect of varying order quantities is shown by the following table:

Order Quantity	Orders per Year	Ordering Cost	Average Stock	Stock Cost	Total Cost
\$ 10	100	\$20	\$ 5	\$1.25	\$21.25
20	50	10	10	2.50	12.50
30	33	6.60	15	3.75	10.35
40	25	5	20	5	10
50	20	4	25	6.25	10.25
60	17	3.40	30	7.50	10.90
70	14	2.80	35	8.75	11.55

For this example the EOQ is:
root(2x0.20x1000/25%), or \$40.

The question of when to order is as important as what and how much. It has already been suggested that a monthly cycle of ordering be adopted, but this does not mean that an order must be placed for each item every month. Whether an order is placed or not will depend on the stock balance in relation to the minimum level calculated for each part.

After an order is placed, it takes a measurable period of time for the goods to arrive. This is known as the "lead time". In an industrialized country where communications are good and the manufacturer is only a short distance away, stock orders can be delivered in a matter of days. In a developing country where parts have to be imported, a lead time of three months is not unusual, and if ports are congested or shipping services infrequent, the lead time may be as much as six months. During that period, there must be sufficient stock to satisfy reasonable demand.

The minimum stock for a part should therefore be that quantity which will cover sales during the period of the order lead time. This can be calculated by multiplying the average monthly sales figure by the lead time, in months.

Assuming that supplies arrive exactly as planned, and assuming that the sales during the period have maintained their calculated average, the last part should be sold just before the new shipment arrives without any customer having been turned away for lack of availability. In fact, however, things often do not work out perfectly. Strikes, storms at sea, payment problems and a host of other factors can affect the time taken to deliver overseas orders. One way of overcoming the problem is to set a deliberately exaggerated lead time but this results in an expensive build up of stocks and defeats the economical order quantity principle.

A better method is to add a safety margin to order quantities. This margin or buffer stock may be calculated as an additional month's sales or as one or two units in the case of high-value items. In the case of low-cost parts where six months or a year's requirements are bought at a time, an alternative is to simply split the stock into two equal portions and place an order when the first half has been used up.

The level at which an order needs to be placed is reached, therefore, when the parts stock is enough to cover the lead time plus the safety factor. It is not necessary however, for all these parts to be actually in stock. Once a regular flow of parts has been created, there will be at least one consignment due to arrive between the time the new order is placed and the time it arrives.

The minimum quantity or "order point" is reached, therefore, when the quantity of parts in stock and already on order (minus any parts owed to customers), reaches the quantity required for sale during the lead time plus the buffer stock.

It is not unusual to set a maximum stock figure in addition to a minimum. If it is found that sales are falling and stocks are accumulating, a temporary halt to orders may be justified to ascertain whether or not the downward trend continues. In this way, an expensive and unjustified investment can be avoided.

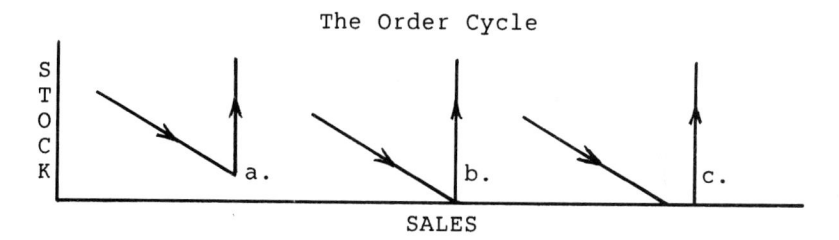

The Order Cycle

```
S |
T |
O |
C |
K |      a.         b.         c.
         SALES
```

a. New stock arrived when there was still a small reserve left in the bin: a good situation.

b. Despite the fact that more stock had been ordered, increased sales meant that the bin was empty just at the point when the new stock arrived: a dangerous situation.

c. Stock actually ran out before the new supply arrived: a situation to be avoided by good planning.

III. MANAGEMENT

The task of management is to anticipate problems that arise continuously and to initiate the changes necessary to resolve them. Good management will resolve problems with minimum delay. Better management will anticipate problems.

Management responsibilities cannot be discharged without authority. The management of any enterprise, whether it be a project, business or government department, needs access to the necessary data and authority to implement changes.

Planning for management starts with the central organization of the project, business or department. The objectives are to produce a crop or provide a service to crop producers if the system is a wholly agricultural one. If it is a transport system, the objectives are to carry a certain tonnage of goods or to provide repairs or maintenance to other transporters.

Sometimes a parts operation has to support two or more distinct user systems. Whatever the circumstances, the organization setting up the parts service should carefully define its objectives. Project objectives are set out in the project document. In a business, the objectives come from the Articles of Association and should be listed in detail each year, or more frequently, in the form of a business plan. Similarly, a government department or ministry should have its objectives defined by the Minister and embodied in its annual budget. The annual objectives are probably part of a long-term strategy for five or more years.

3.1 Planning objectives

Management must plan the means to achieve the objectives of the enterprise. If a given number of machines are to be kept running efficiently and economically, then parts will have to be provided. In order for the parts to be available when needed, stocks have to be established close to the machines. Stocks will then have to be replenished in a regular and well-organized manner.

A parts operation may be commercial or non-commercial but even a non-commercial operation should run as though a nominal profit is required. Otherwise it will run at a loss and have to be subsidized by some other part of the system. Profit is not simply an excess of income over expenses which can be handed out to shareholders. It has a variety of functions vital to the continuity of the business. Profits should go into a reserve fund from which new business investments can be made and from which the cost of writing off stock and equipment depreciation can be supported.

Management planning has to take into account the physical requirements of the parts operation; its stores, its staff and its funds. The parts department can be a key to the success or failure of the enterprise and a proper degree of importance should be given to its needs.

3.2 Continuity of supply

In an agricultural project, parts may be delivered with the new machines at the start of the project. Similarly, agricultural ministries sometimes buy parts mainly as an adjunct to a new machine order in the expectation that they can cater for the older machines in the fleet. This method does not keep the machines running for very long. First, some units inevitably suffer abnormal failures from accidents or clumsy handling and will require parts that were not in the initial stock order. Second, parts in the initial stock will be used at a variable rate which can be hard to anticipate. These parts need to be replaced into stock at intervals which may not be regular.

The system described above is not a substitute for a properly planned parts supply programme. The establishment and management of an efficient parts flow from manufacturer to user is the most important function of the parts department. Organization and care of stock is secondary.

For the commercial machinery distributor the task of keeping many customers supplied with parts can only be achieved through a well-organized parts department. Thought must be given to planning so that the right parts are available where and when they are needed and at the right price.

Manufacturers can help identify the parts that will probably be needed in the early stages of machinery operation. Once the equipment starts to operate, a different picture will appear. Operating conditions vary from territory to territory; even from farm to farm. There are differences in soils, in the quality of fuels and lubricants. Damage from rocks or stumps in the soil varies. The quality of drivers, their training and remuneration differs. The quality of operator supervision has a bearing on operator habits and adherence to proper maintenance schedules. All these factors impinge on the rate of parts usage and the type of parts that will be required.

It is essential, therefore, that these requirements are effectively taken care of. Accurate records of parts movements must be built up. Rapid action must be taken to acquire and deliver parts that are not in stock when needed. These details must be recorded and analyzed piece by piece and day by day so that stock can be brought closer to meeting the needs of the equipment in operation, and so that these requirements can be anticipated for the months and years ahead.

Planning several months ahead is necessary for stock orders and planning years ahead is advisable for expanding and ageing local machinery populations since more storage space may be needed and more staff may have to be trained.

3.3 Planning details

Because the parts operation seems complex, it is sometimes relegated to a simple storekeeping function and its proper development is ignored. To avoid a system breakdown, it is worthwhile spending time planning the operation and finding and training the right people.

The parts department plan should be carefully structured under the headings suggested below. The emphasis in this chapter is on the necessity for planning, for the staff to have authority within the plan and for controls to ensure that such authority is not abused.

The parts department plan should cover:
- stocks - machine populations to be covered
  - turnover rate
  - service level
- stock control
- stock re-ordering
- emergency procedures
- general operating procedures
- staff requirements - numbers
  - training
- costs
- overall budget
- reporting procedures
- management reviews

3.4 Responsibility and authority

The parts department manager must have the authority to take all the day to day decisions and must be able to delegate responsibilities to subordinates.

Delegation of responsibility is particularly important in a parts operation where a large number of small decisions must be taken. It is difficult for those outside to be close enough to the myriad of details to take responsibility. For the upper levels of management it would be excessively time consuming.

The parts department plan should also detail the control structure by which the results of the day-to-day decisions can be monitored. Here the management of the organization or project can keep overall control of the department. Variance from the plan can be seen immediately and the reasons demanded. The parts manager can be instructed to bring the department back in line with the plan. Or, in consultation with the parts manager, the plan can be modified to take account of changing circumstances.

Frequent monitoring is advisable where the less senior members of staff make important decisions such as the commitment of funds to ordering parts.

3.5 Control reporting

Regular reports enable senior management to monitor the parts operation. Accounting and clerical staff should present reports comparing performance with the plan. Monthly reporting is necessary for close control. Reporting over a longer period may allow laxity and catching up can be difficult and expensive.

One set of bad reports should be investigated but there may be a simple isolated reason. When successive reports show an adverse trend, however, action should be taken to return the operation to its planned course.

Monthly reporting should not be too onorous otherwise quality will slip. Levels of stock, movement and customer service must be monitored in the context of operational cost.

3.6 Stock level control

Stock level is simply computed using a Stock Report (see page 47). Opening stock is, initially, the stock with which the department started business. Then, month by month, it becomes the Closing stock figure from the previous month.

The Month's sales column indicates the turnover rate and is derived by dividing the Closing stock figure by the average monthly value of the previous twelve months Cost of parts out. A stock representing six months sales turns over twice per year. The

Stock Report

Date	Opening Stock	+Cost of Parts in	-Cost of Parts out	Closing Stock	Month Sales	Planned Stock

Planned stock level indicates progress in relation to the plan. If the Closing stock greatly exceeds the planned level, action will have to be taken to reduce it. This may mean a special programme to promote sales, a request to the workshop to consider an early overhaul programme, or a request to the financial overseers to write off dead stock. This should not arise early in an operation but, for good accounting, it should be carried out annually. Stock that does not move is an over-valued asset, and distorts the operation's financial position.

If stock is seriously below planned levels, an explanation is equally important. Are there delays in the arrival of orders or has there been a breakdown in the parts ordering process? The Planned stock level is based on equipment needs, and low stocks will mean poorly serviced equipment.

3.7 Movement reporting

The level of movement should be reported by using a Movement Report (see page 49). The first part of the report shows how movement has progressed in relation to the plan. If movement is seriously below plan, in an agricultural project it may mean either that service schedules are not being maintained or, less likely, that the equipment is behaving better than anticipated. If sales are above plan, there may be a case for increasing stock, but this will be determined also by the lost sales and service level reports. It may also point to mechanical weaknesses in the equipment which should trigger an investigation and action from the manufacturer.

The last five columns in the movement report indicate characteristics which may be used for administration planning. For instance, if there is a large number of small orders from a project workshop, it may mean lack of systematic job preparation in that department. Without improvement it could mean excessive work for the clerks and possibly a need to increase staff. A small number of orders may indicate that present staff are under-occupied and that some savings could be made.

Movement Report

Date	Month Sales	Total YTD	Plan YTD	Var. %	Ords No.	Items No.	Ord $	Item $	Item/ Ord

3.8 Service level reporting

The Service Level Report (see page 51) should indicate the relationship between the number and value of items called for and the number actually provided at first call.

Each time a request is made of the parts department, a customer order form should be compiled. When the parts are supplied, a copy of this form will be the invoice, another copy the delivery note and a third copy the accounting document. When the parts are not supplied the document will be used to signal back-order procedures (see Chapter 2).

The Items supplied column comes from the sales report. The Not supplied column is an analysis of the customer orders not invoiced. It should distinguish between stock items (S) and items not normally carried in stock (NS). Failure to supply stock items shows inefficiency. A demand for a significant quantity or value of non-stock items may call for a wider stock range, especially where significant values are involved.

The true service level is the relationship between items supplied and those requested:

$$\text{COLUMN C} = \frac{\text{COLUMN A} \times 100}{\text{COLUMNS A + B}}$$

Any continuing divergence from the planned level needs investigation. Reports above plan indicate either excessive inventory investment or naive and inaccurate reporting. Reports below plan imply reduced equipment availability.

Service Level Report

A. Items Supplied		B. Not Supplied				C. Service		
		No.		Value				
No.	Value	S.	N.S.	S.	N.S.	No.	Value	Plan Service Level

3.9 Back order reporting

The Back Order Report (see page 53) records the items that were not supplied on first demand, but delivered later. While they do not change the true service level, they indicate how efficient the back order service has become. If column C consistently records supply of 85 percent of the items required, and the back order report consistently records an additional 15 percent of the items supplied, then the overall effective service level is 97.75 percent (85 + (15 percent of 85)), and the remaining 2.25 percent can be regarded as lost sales.

The average age of back orders when invoiced will give a measure of the efficiency of the overall inventory control system. If it is the same as the time taken to acquire goods from the manufacturer on special order (the emergency order lead time), then order levels could be increased. If the average age of a back order is less than the emergency order lead time, then back orders are being filled from stock orders in the pipeline, and an increase in minimum stocks can improve the level of service.

The Trading Summary Report (see page 54) shows the relationship between the income and expenses. It compares cumulative year to date (YTD) figures against those planned for the same period. This gives a clearer yearly picture since monthly figures may contain misleading anomalies.

Even where a profit is not required, for example where the parts operation serves its own government department or an externally-funded project, the same principles apply. The relationship between expenses and volume of parts handled is important and the relationship between costs incurred and planned expenditure is even more important, since some types of corrective action (price increases for example), are not possible. If the parts department is a profit centre in the original plan, its costs can be carried forward to the next level of operations (for example, cultivation or transport), which can in turn pass on their costs so that the result is a clear cut figure for the real cost of producing goods or services.

Back Order Report

Back Orders Supplied		% of Total Items	
No. of Items	Value	No.	Value

Trading Summary Report

Sales		Gross Profit			Expenses			Net Profit		
Value	Cost	M	YTD	Plan	M	YTD	Plan	M	YTD	Plan

3.10 Order control

Orders prepared by staff have to be checked by the parts manager before signature. Management has to be sure that the order is justified and have some idea of its value, even without an official proforma invoice. If an order is compiled on an internal document called a Stock Replacement Indent (see page 56), details can be shown to justify and cost it.

The stock replacement indent should show stock type, description and part number, movements over the past four quarters and the calculation of order level and economical order quantity (or the category from the guide chart). The current stock balance should also be shown.

The lead time determines how much is kept in stock and on order. Any delay in checking and approving the indent will lengthen the lead time. The ordering process should be kept to a rigid cycle, and be completed expeditiously.

A problem arises when parts are needed that have not been stocked previously. The item may be procured locally or it may need to be treated as an emergency order. Often a snap decision is taken to buy two instead of one in case it is needed again but if the demand was an isolated occurrence, the second unit will join the dead stock.

In some computerized inventory control programmes, new stock can be tested before being automatically re-ordered. It is usually wiser to wait until the second demand and then buy two if the part falls into the low to medium value category. Just as new parts come into demand, so established items will eventually drop out. At first, demand may slacken as the active machinery population diminishes. Eventually these parts will be required only rarely and then not at all. Other parts may drop out of use quite suddenly if they are replaced by an improved version. The un-needed parts then represent an investment with no further earning potential.

Stock Replacement Indent Form

Stock Type						Lead Time	Mo.		Date		
Desc.	Part No.	Sales				Ord Lvl	Act Bal	Ord Qty	To Order		
		Q1	Q2	Q3	Q4	Av				Qty	Price	$

Regulations about what may be written off vary between countries and organizations. In some cases, a board of enquiry has to be satisfied that there is no further value in the item, and that it is physically destroyed to prevent possible fraud. In other cases, it is the tax authorities who have to be satisfied that a portion of the company's gross profit should not be taxed because it has been used to cover the loss of what had been thought of as assets.

A depreciation allowance should be made in the annual plan to cover a proportion of stock write-off, because only then can the inventory be properly valued and a realistic turnover rate measured. The financial controller needs to assess the year's activity before making a final decision.

Where official distributors are concerned, a stock write-off policy can be a powerful stimulus to stocking. Some manufacturers accept and credit surplus stock returned to them at full invoice value. Sometimes only certain categories can be returned. Other categories may be written off and credited at a lower value. In most cases stock returned has to be in good saleable condition and in its original packing.

Stock return policies do not normally apply to buyers other than the distributor. If a project or a government department buys directly from the manufacturer, there is normally no method of recuperating the value of surplus stock unless another private buyer can be found for it. This is a further reason for leaving it to the commercial distributor to carry the largest share of the stock holding burden.

3.11 Reasons for budgeting

A budget enables resource allocation to be planned so that estimated expenses can be met from estimated income. Where income is not expected to arrive in time to meet expenses, a clearly-defined budget helps in arranging a loan to bridge the gap.

3.12 Budget headings - income

In a government department or an agricultural project, there may be only one source of income: the government treasury or a banking arrangement designated to handle treasury funds. Government departments and projects are sometimes set up to receive other income from, for example, the sale of crops or agricultural inputs such as fertilizers and agricultural chemicals. Sometimes machinery, parts or tools are sold to provide income for a department or project. All sources of income should be included when setting out the budgetary income statement.

Income receipts may not be regular. Government funding usually follows publication of an annual budget, but there may be some delay to allow for discussion and allocation. Even when funds are allocated and become available, they may be subject to periodic limits. For example, in some countries, local treasuries have a restricted monthly cash limit. Cash is available on a first come, first served basis, and the departments with the clearest income budgets are first in the queue.

In a commercial enterprise, the income of the parts department may come from a variety of sources including the enterprise's central treasury. It is most likely to come entirely from parts sales and, if the enterprise is successful, it will more than cover the department's expenditures.

Commercial parts sales connected with agriculture are still likely to suffer seasonal fluctuations. An attractive policy is to promote parts sales in the seasons of low farming activity. This evens out cash flow and also encourages the machinery user to overhaul equipment during quiet periods in preparation for the busy cultivation and harvest seasons. However, most machinery is still left until it is needed, and consequently there is a heavy demand for parts during the busy season.

Seasonal income fluctuations must be considered in the budget. Even in a successful commercial enterprise, there may be months when income is below expenditure and procedures for borrowing from the central fund must be established in advance.

The annual business plan is usually sub-divided by month and income and expenditure budgets should be similarly sub-divided as wages and bills are likely to be paid on a monthly basis. Some services may be billed on a quarterly basis so it is also useful to re-group the monthly figures into quarterly summaries.

Where sales are made on credit, the income may not be received regularly at the end of the following month. Patterns of credit recovery vary.

If cash sales are low there may be ways of stimulating demand with special offers or promotions. In periods of economic austerity, the most significant drop may be in credit sales income, because customers are paying more slowly. They usually take maximum advantage of supplier credit. There are two types of income figures to watch, the cash received daily and monthly, compared with planned cash flow in the control reports. The second set of data to be reviewed is the information relating to outstanding accounts.

Accounts are normally paid on a 30-day basis which should mean that 30 days after the issue of an invoice, the funds are received. In practice, this does not always happen. It is more usual for the customer to wait at least until the end of the following month, when a statement of account is received. If parts are bought early in the month, this amounts to almost two months credit. Some customers hold back payments on the assumption that their business is sufficiently important for the practice to be acceptable.

The usual method of reporting outstanding bills situation is to classify the bills by age: those not overdue, those that are up to 30 days overdue, then 30-60 days, 60-90 days, and finally those more than 90 days overdue. The organization's financial director should see this report every month. Overdue bills represent money which should be re-invested in stock where it will earn profits, and therefore represent a loss.

Attempts to charge penalty interest on overdue accounts seldom succeed, and while the payment of a bill may be enforceable under law, this does not always apply to interest charged on it. To encourage prompt payment, the invoice should make it clear when payment is expected. A monthly statement should be sent to each customer with an outstanding account, not only as a reminder, but also to provide a vehicle for the regular reconciliation of accounts.

Statements should be sent regularly, starting with the month the invoice is dated, even if it is not yet due for payment. They should arrive on the customer's desk just before the end of the month, so that they coincide with month-end accounting when they are more likely to be paid. The statement should show all invoices raised and payments received and the overdue amounts should be identified by age as a further reminder. When payments are not being received on time, attention can be drawn to this by adding warning messages on the statement.

The enterprise should demonstrate that it expects prompt payment, and will take action if necessary.

Credit-stopping action should be taken as soon as there are signs of any customer, no matter how important, falling behind in payments. If two consecutive statements fail to show receipts, or if outstandings move into the 90-day column, firm action should be taken. The ultimate resort of going to court to recover large sums is costly, however, and may be a slow process.

The monthly receivables report may show a total close to the average monthly sales if credit control is good and there is a high proportion of cash sales. It may represent two or three month's sales if most business is done on credit. If receivables add up to more than three month's sales, cash flow will be under undue stress and action should be taken. By this time, an excessive number of accounts will have moved into the 90-day column and customers may believe they can delay payment indefinitely.

Treatment of cash sales must be controlled. Procedures should be developed to ensure that cash handed over to the treasury department corresponds to invoices written, and that this is reconciled on a daily basis. The monthly invoices should be totalled and presented as a sales report. These figures will be broken down to provide the various components relating to outward movement of parts in the management reports.

The invoices must be posted daily to the stock cards. At the same time, the cost of each part must be entered on the invoice and totalled. These figures provide the accounts clerks with the sales figures and the costs, and by subtracting one from the other, the gross profit. In a commercial organization the gross profit figure is more important than the sales total, since it is from the gross profit that expenses have to be met.

If a parts-using (rather than a parts-selling) organization, such as a project or government department is set up as a series of profit centres, the same control systems will be needed, even though the objective may simply be to pass on the true cost of operating to the next department.

3.13 Pricing

Profit on sales depends on the price charged in relation to cost. Price control is, therefore, an important part of the system. The price charged for a part is determined by several factors:

- manufacturer's list price;
- agreed discount due to the purchaser;
- additional discount for special promotions;
- any premiums charged for emergency status;
- ancillary packing and local handling charges.

Unless prices are "free on board" (FOB), additional charges will be made for:

- freight;
- insurance;
- port charges at destination.

To arrive at the total "in-store" cost, the following must be added:

- customs duties;
- customs handling fees, stamp duties and overtime;
- transport charges from port to store.

To the "in-store" cost a mark-up is added to cover:
- depreciation;
- rent (the space costs for storage);
- insurance;
- staff costs;
- lighting and heating;
- postage, stationery, telephone and telex;
- other incidental overheads including charge for central accounting and management facilities;
- financing costs;
- eventual net profit (which should at least equal the interest that could be earned by the capital employed if it were put to other uses).

A part costing $10 (net) from the supplier, could be shown on a customer invoice at a retail price of $24, or more, if high levels of duty have to be paid.

The calculation can be broken down as follows:

```
 cost FOB.                    $10.00
+freight and insurance          1.50   (15%)
=landed cost                   11.50
+duty                           1.73   (15%)
+handling                       0.57   ( 5%)
=total in store                13.80
+depreciation                   5.52   (40%)
+other overheads                2.07   (15%)
+finance (inc. net profit)      2.07   (15%)
=retail price                  23.46 (mark up 104% on
                                      landed cost)
```

If the depreciation charge looks high, it has to be remembered that 80 percent of parts move so slowly that they account for only 20 percent of sales. Considerable allowance has to be made, therefore, for dead stock to be written off. This still applies where good turnover rates are achieved.

Management control over prices is important because there is a tendency for costs to become inflated while prices become eroded. For example, if discounts are combined with credit abuse by many customers, gross profits can easily slip below the level of normal expenses. Price discounts should be accorded only to customers who qualify in terms of sales volume and prompt payment.

The final prices are calculated by the entry clerks as they post invoices for incoming goods and are written onto the stock cards. The simplest way of making the calculation is to apply a standard factor to the manufacturer's invoiced price. Thus, where the manufacturer's price is $10.00, the factor used would be 2.346 to reach a retail price of $23.46. The effect of differing currency exchange rates can be built into the cost factor and checked frequently.

When a new consignment arrives, there may be parts in the bin at an old price and the cost and retail price of the new stock may be significantly higher. There are three ways of dealing with this situation. The first is to mark all the stock at the new price. The effect of overcharging for some items is balanced by unrealistically low prices from slow moving stocks which have not recently been replenished.

The second method is to sell the old stock at the old price until it is exhausted. This is the most equitable method, but it entails duplication of stock cards and can cause confusion; there should only be one card per part number.

The third method is to average the old and new prices but this requires more calculation by the entry clerks, for example,

```
Old stock:    12 items at $21.20 =   $254.50    (old
                                                price)
New stock:    50 items at $23.46 = $1 173.00    (new
                                                price)
Total stock: 62 items at $23.02 = $1 427.40
                                      (averaged price).
Price to be changed on card from $21.20 to $23.02.
```

Management must ensure that the rules on pricing are clearly understood by the staff, and that checks are made to see that they are applied.

The standard factor method is the simplest in practice, but it requires frequent checking to take into consideration currency fluctuations, and duty rates and transport charges which may be subject to exaggerated inflation.

Parts are sold in competition with other sources of the same or similar items. If unreasonably high prices are charged, customers will look elsewhere, not only for their parts but also for their new machines. Control may be exercised by reducing prices on parts which are available from competitors and adding to the price of captive parts for which there are no other sources.

Income control ensures that all income is properly accounted for, that sales on credit are realized as cash income, and that income opportunities are not missed by inefficient pricing or marketing.

3.14 Budget headings - expenditure

Budget headings for expenditure will vary from place to place, but they will always have some elements in common:

- wages and salaries;
- electricity for lighting and heating;
- other fuel costs for heating;
- telephone (and possibly telex);

- postage;
- stationery;
- maintenance;
- transport.

Control over expenses requires constant vigilance, although in a period of inflation it may be possible to forecast the escalation of costs when preparing the budget.

As in the case of monthly income, monthly expenses should be efficiently accounted for so that monthly reports reflect the true picture and either satisfy management that the plan is being followed, or trigger corrective action to bring the operation back to planned objectives.

3.15 Salaries

The most important component of an expenditure budget in any organization is weekly or monthly payments to the people who work in it. Staff may be expanded or diminished, but provisions must be made to pay the staff engaged in the organization at any given time on a precise schedule.

The calculations needed are not complicated. They include the basic pay, according to each member's contractual pay scale, plus any overtime, taxes, insurance (contributory, statutory or otherwise), other allowances and contributions to health or social schemes.

Since these all have statutory or other legal foundations, failure to meet any of them can be subject to eventual claims, or more stringent legal action. In Chapter 4 there are suggestions on how to properly take care of these elements. Normally, salary expenses can be planned months ahead or for a whole year at a time.

Variable amounts of casual labour or overtime may be needed - usually on an hourly basis. Wherever possible, these costs should be anticipated and provision made in advance. Stock-taking, for example, or a major reorganization of parts resulting from building operations, can be planned ahead.

If experience proves that there are often unforeseen additional labour costs, then a regular reserve should be added to the monthly wage budget, calculated perhaps as a percentage of the norm. Variable amounts are usually due only to weekly paid staff, and casual labour or overtime can be controlled closely enough over a month to make it conform to a monthly budget.

3.16 Other expenses

The next five headings in the list are normal office expenses. At the beginning of a project or when an enterprise is started up, such costs have to be estimated. Some, such as heating and lighting, can be calculated, given the size of the premises. Others, particularly communications costs, have to be judged on personal experience or advice from, for example, accountants who have access to figures for similar businesses in the same locality.

Initially, budgetary provision for communication costs - postage, telephone and telex - should be generous; it will give the overall plan a more realistic aspect and the budget can be trimmed subsequently in the light of experience. Communication is essential - too often lack of it causes bottlenecks and delays in execution of a project. Communication costs need strict control since they are open to abuse. The parts manager should verify each charge on an itemized telephone or telex bill. This often imposes caution. Personal telephone calls should be discouraged.

Maintenance is often neglected in the budget. The personnel costs of keeping the parts department in good order are covered by staffing provisions, but materials are required for cleaning and for keeping the buildings, furniture and fixtures, fittings and equipment in good condition. Initially, the cost has to be estimated. Subsequently, budgetary provision should be judged on the basis of past costs and the current state of the items to be maintained. If maintenance is neglected in one year, it may cost more than twice as much to bring the facilities back into good condition the next year, and in the meantime the work of the department will deteriorate, along with its surroundings.

Promotion of the parts department's activities is usually only required in a commercial situation. It should be considered part of the organization's overall promotional expenses, since there should be close co-ordination between promotion of machine sales and any exercise to stimulate parts sales.

Provision should be made for transport under three sub-headings. First, there is a need to fetch parts from the major supply source on a regular basis. For an importing organization, this is usually from the docks or airport. Parts also need to be delivered to customers. Whether these services are provided by the organization's own transport or by hired transport, they must be budgeted.

Second, shopping locally for items not in stock should be considered separately since it is often time consuming and if ignored it can become a function of the workshop and its highly-paid mechanics.

Third, transport for emergency orders must be budgeted. In balancing inventory costs against the cost of emergency orders, freight is usually a major consideration. Sometimes it can be passed on to the consumer, but in general this is not the case.

3.17 Purchase control

Cash purchases of supplies or parts need strict control. Vouchers for drawing cash are commonly used and the parts manager should have authority to draw up to a certain amount. Above that figure, it should be authorized and signed by the financial director of the enterprise.

Discretion in local purchases may produce economies, considering the range of prices found by a little research and also the cost of time and transport. Control over local purchases is also needed in cash-flow planning. It makes no sense to negotiate favourable prices or generous discounts with local suppliers, only to lose the facility because funds cannot be found to pay their bills on time.

A driver may be able to pick up parts or fuel by signing a delivery note or an unpriced invoice. Where this is a feature of local purchases, the price must be obtained immediately without waiting for the supplier's invoice to arrive. Otherwise purchase and expense budgets cannot be controlled with any precision and parts sold to third parties cannot be priced immediately. This also applies to parts bought for the organization's own service workshop where job pricing must be kept up to date.

Cost control should also be applied to purchases from the principal manufacturers serving the organization. Bulk purchase reduces ordering costs more than the additional stocking cost. Special discounts or promotions may be taken advantage of, and the parts manager should know of their existence.

Cost control should not be simply a means of staying within the budget. If ways can be found to reduce costs below budget levels, surplus funds can be put to good use in bonuses for staff or service improvements.

IV. STAFF

4.1 Organization

Staffing a parts department begins with identification of posts to be filled. An organizational chart, such as the one shown on page 70, outlines the internal structure of the parts department and its relationship to the rest of the organization. Some of the posts are essential, common to parts departments anywhere. Others depend on local circumstances such as the availability of services rendered by other departments of a central administrative section in the organization. This diagram is therefore not the only possible organizational structure; specific situations may require other patterns. Any organizational chart should be regarded as a guide - a non-rigid framework which shows approximate relationships between people and functions. It is also a useful tool to show new employees where they fit into the organization.

While posts in a smaller store may be combined, three basic functions - store supervision, attending to customers's needs, and clerical administration - should be regarded as separate. Even if the store is so small that all jobs are done by a single person, the duties of that person should be stated in terms of the separate functions.

In a larger framework, experienced employees, particularly those with special skills, are less easy to fit into a formal structure and attempts to do so can result in frustration and mismanagement. Skilled managers understand how workers function best and organize accordingly to maximize benefits.

Some basic principles of organization are worth emphasizing. First, each member of the organization should have only one direct supervisor to avoid any chance of conflict in the instructions he or she re-

Organizational Chart

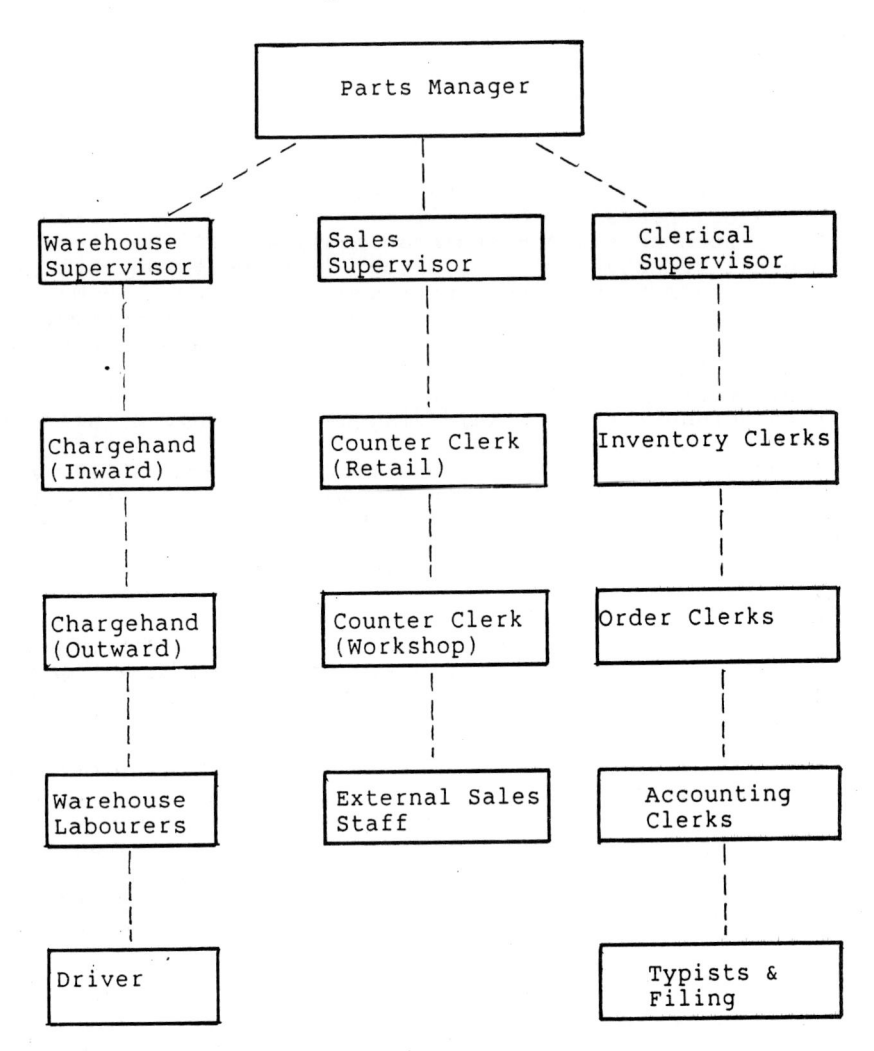

```
                        ┌──────────────────┐
                        │  Parts Manager   │
                        └──────────────────┘
              ┌──────────────┼──────────────┐
    ┌──────────────┐  ┌──────────────┐  ┌──────────────┐
    │  Warehouse   │  │    Sales     │  │   Clerical   │
    │  Supervisor  │  │  Supervisor  │  │  Supervisor  │
    └──────────────┘  └──────────────┘  └──────────────┘
            │                 │                 │
    ┌──────────────┐  ┌──────────────┐  ┌──────────────────┐
    │  Chargehand  │  │ Counter Clerk│  │ Inventory Clerks │
    │   (Inward)   │  │   (Retail)   │  │                  │
    └──────────────┘  └──────────────┘  └──────────────────┘
            │                 │                 │
    ┌──────────────┐  ┌──────────────┐  ┌──────────────────┐
    │  Chargehand  │  │ Counter Clerk│  │   Order Clerks   │
    │  (Outward)   │  │  (Workshop)  │  │                  │
    └──────────────┘  └──────────────┘  └──────────────────┘
            │                 │                 │
    ┌──────────────┐  ┌──────────────┐  ┌──────────────────┐
    │  Warehouse   │  │External Sales│  │    Accounting    │
    │  Labourers   │  │    Staff     │  │     Clerks       │
    └──────────────┘  └──────────────┘  └──────────────────┘
            │                                   │
    ┌──────────────┐                    ┌──────────────────┐
    │    Driver    │                    │    Typists &     │
    │              │                    │     Filing       │
    └──────────────┘                    └──────────────────┘
```

ceives. The line of authority from manager through the various levels of subordinate staff should be clearly defined.

Second, the objectives of the operation should be clear and made known to all staff so that they understand why each function must be performed and how it relates to the others.

Third, while a clear definition of authority and responsibility for each task is basic, success also depends on smooth internal relationships between people and jobs at each level of the organization. For example, the clerical and warehouse staff depend on each other at many points in the system for the services and information they need to function, although neither is dependent on the other for authority. The lateral relationships are important too.

It should be emphasized that all members of the organization are mutually interdependent, and that cooperation is the only way to ensure success. This is true within the department and also in reference to relationships with other departments. At the same time, all staff should recognize that others have specific jobs to do, as defined by the chart, and that mutual cooperation maximizes efficiency.

The fourth point to emphasize is that responsibility for any job must carry with it the necessary authority. This also means a parallel limitation on responsibility which must be taken into account when planning the organization's development.

4.2 Staff responsibilities

The responsibilities of each member of the staff, including the manager, are best defined by a job description. this should both detail day-to-day tasks and also clarify who each staff member takes orders from and the nature and extent of their authority over others. It should refer to lateral relationships and indicate other functions from which and to which services should be made available, In addition, the job description should inform the employees what constitutes a standard of satisfactory performance and what proof of this must be made available to their supervisor.

The job description is also useful when hiring replacement staff. It allows the manager to closely define what sort of person is required and to compare an applicant's capabilities with job specifications.

It is essential that each staff member has a clear idea of the scope he of she has for making decisions, particularly when money is involved. Normally, limits are placed on cash handling which vary according to the seniority of the employee. Guidelines for handling cash must be laid down precisely in a procedural manual rather than in the job description, but the job description may refer to the range of responsibilities involved in the position.

Other financial decisions, particularly those regarding the expenditure of money, may be included in the job description. In the case of the parts manager, limits may be imposed across his entire range of responsibilities, including purchasing stock. The authority to commit funds to stock is an important part of the manager's job. It must be clearly understood that this authority applies equally to local as well as overseas purchases.

The job descriptions and profiles that follow are based on the organizational diagram at the beginning of this chapter. They give some guidance as to the skills, responsibilities and characteristics required for posts in a parts department. No profile will exactly match any one of the applicants for a post, but the job profiles do provide a standard of comparison.

The job descriptions for junior staff are much more detailed than those for senior staff because senior staff members are expected to be more aware of the policy objectives of the organization and to be able to use their own initiative in finding ways to achieve these objectives. Junior staff members need more guidance and the job description offers a good framework for a perpetual training programme.

4.3 Parts manager

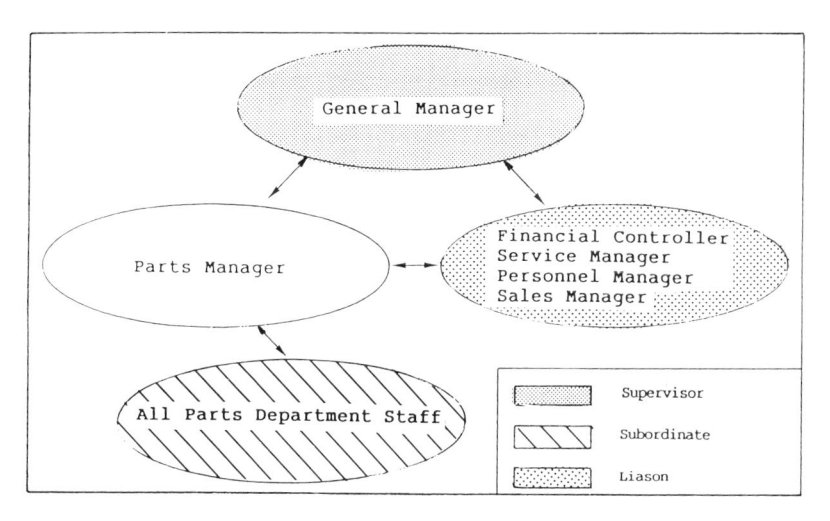

```
                    General Manager

  Parts Manager  ←→  Financial Controller
                     Service Manager
                     Personnel Manager
                     Sales Manager

  All Parts Department Staff    ▓▓▓ Supervisor
                                ∕∕∕ Subordinate
                                ▓▓▓ Liason
```

Reports to: General manager (or Managing director).

Summary of responsibilities: The parts manager is responsible for the administration of the parts department and for the achievement of performance targets set by the general manager within the budgetary framework established for the department.

Detailed job description: The parts manager is responsible for the overall day-to-day control of the parts department, its stock and its staff. The parts manager is specifically responsible for carrying out, or delegating to subordinate staff, the following tasks:

  continuous review of the parts requirements of the machinery population served by the parts store;

  selection, in consultation with both manufacturers and users of the machinery, of the range of parts to be stocked;

preparation of orders for parts to be purchased, and finalization, in consultation with the financial controller, of all arrangements for payment and shipment of orders;

planning of suitable parts storage facilities and arranging for such facilities to be available as required;

supervision of the receipt and storage of parts and their registration on the stock card index;

supervision of the issue of parts to the satisfaction of the machinery users;

procurement of parts not in stock and their delivery to the users;

supervision of the pricing and invoicing of parts;

supervision of the maintenance of up-to-date sales records, income, and outstanding bills;

arranging for collection of payments on outstanding bills;

supervision of the registration of all sales on the stock cards;

planning of stock changes to take account of future customer requirements;

supervision of the arrangements for protecting the stock from theft, deterioration or loss;

appointing staff to operate the department within the organizational and salary structures agreed with management;

arranging for staff to be trained in the various skills required to operate the department;

arranging for staff to be kept informed of the policies of the enterprise and its suppliers;

supervision of the application of the rules of the organization;

resolution of staff problems and grievances;

termination of the employment of staff as and when necessary, as defined by the general rules and procedures of the organization, and in accordance with any existing labour union agreement or employment legislation;

preparation of and delivery to senior management of the reports and statistics required to evaluate operation of the parts department.

Authority: The parts manager has authority to engage and dismiss Parts Department staff in accordance with agreed budgets and procedures. The parts manager has authority to commit funds for expenditure within the agreed budget, but with the following limits on individual cases:

any cash expenses in excess of $...... per month to have prior agreement from the financial controller;

any individual local purchases in excess of $...... to be agreed with the financial controller in advance;

any individual parts orders valued in excess of $...... to be agreed with the financial controller before dispatch;

any capital expenditure to have prior agreement from the financial controller. The parts manager has authority to offer credit to customers for orders up to a value of .....$ or up to credit limits set by the financial controller.

Liaison: The parts manager works with:

the financial controller on all matters relating to budgets, expenditures, and credit control;

the service manager on all matters relating to the provision of parts to the service department;

the personnel manager on staff matters;

the sales manager on questions of support to equipment in the field.

Standards: The parts manager's performance will be judged by the performance of the parts department in relation to planned goals, as reported in the monthly accounts. Particular importance will be attached to gross profit achieved in relation to direct costs, stock turnover and customer satisfaction.

Parts manager: Profile

Formal qualifications: The parts manager should have a formal education indicating a satisfactory standard of literacy and numeracy. In an importing organization, some knowledge of the principal languages of the supplying countries is an advantage. Education beyond high school is only necessary for larger organizations where a polytechnic diploma in engineering or a university degree may be required. Training in administration would also be an asset.

Technical training: The parts manager should have had some technical training, either in a polytechnic or vocational training institute, or through suppliers of parts of equipment. Some technical knowledge of the principles of machinery construction and operation is necessary.

Experience: The parts manager should have at least five years previous experience working in a storekeeping environment. For a larger organization, the parts manager should have previous parts management experience, or at least have held a senior position in a parts store. For a store with several employees, the parts manager should have previous experience in parts sales, purchasing, and inventory-control techniques.

General character: The parts manager should be accustomed to dealing with people in a machinery-oriented environment, and be interested in resolving their problems. Honesty, tact and energy in carrying out the job are all essential requirements.

4.4 Warehouse supervisor

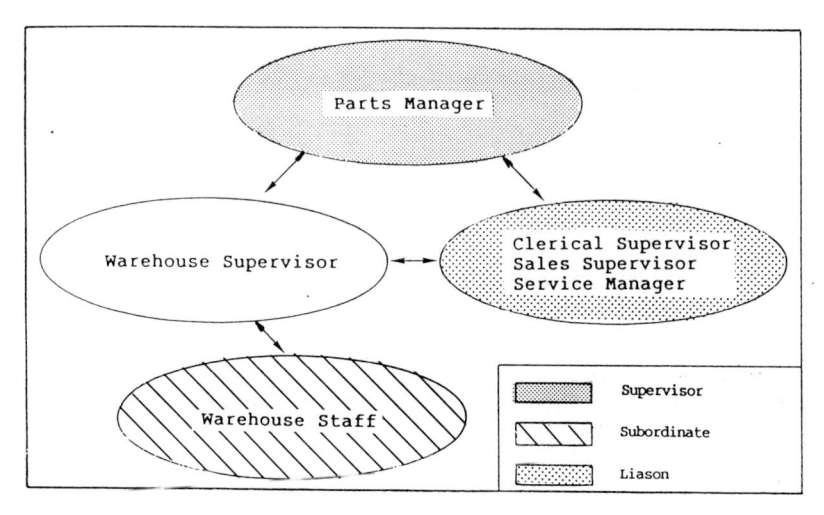

```
                    Parts Manager

Warehouse Supervisor  ←→  Clerical Supervisor
                          Sales Supervisor
        ↓                 Service Manager

   Warehouse Staff

        ▓▓▓  Supervisor
        ///  Subordinate
        ░░░  Liason
```

Reports to: The parts manager.

Summary of responsibilities: The warehouse supervisor is responsible for the efficient operation of the warehouse, including security, condition of parts and their orderly and timely delivery.

Detailed job description: The warehouse supervisor has the day-to-day responsibility for the following tasks:

supervision of receiving, checking and proper storage of all incoming parts;

immediate written notification to the clerical department of any damage, discrepancies or shortages in incoming parts;

written notification to the clerical department of any changes in parts location;

supervision of the daily cleaning of the floors, shelves and other fixtures in the store and the removal of rubbish;

supervision of orderly placement and clean condition of parts;

supervision of parts selection, according to orders received from the sales department, and the assembly of orders for delivery to the sales department;

notification to the sales and clerical departments of any shortages not apparent from stock records;

supervision of packing of orders for delivery to customers;

supervision of parts transportation when the department operates its own vehicles, or arrangement of delivery through external transportation;

supervision of staff training on a day-to-day basis in accordance with staff development programmes;

preparation of annual staff reviews.

Authority: The warehouse supervisor has authority over the day-to-day activities of warehouse staff.

Liaison: The warehouse supervisor liaises with the clerical and sales supervisors for day-to-day exchange of information necessary to the operation of the department, and with the service manager regarding delivery of parts to the workshop and local purchase of non-stocked items or consumable supplies.

Standards: The warehouse supervisor will have achieved a satisfactory standard when:

received parts are all correctly binned;

the clerical department has been notified in writing of all discrepancies on the day that they were discovered;

the warehouse is clean and tidy;

parts orders are accurately delivered and checked within a reasonable time;

staff reviews show that progress in training is being maintained according to the established programme.

Warehouse supervisor: Profile

Formal qualifications: The warehouse supervisor must be educated to a reasonable standard of literacy and numeracy.

Technical training: The warehouse supervisor does not normally require formal technical qualifications but some technical knowledge is beneficial for parts identification. This may be provided within the organization or by suppliers.

Experience: The warehouse supervisor should have at least three years experience with a machinery maintenance and repair organization.

General character: The warehouse supervisor should be a tidy person with an instinct for order, discipline and cleanliness.

4.5 Clerical supervisor

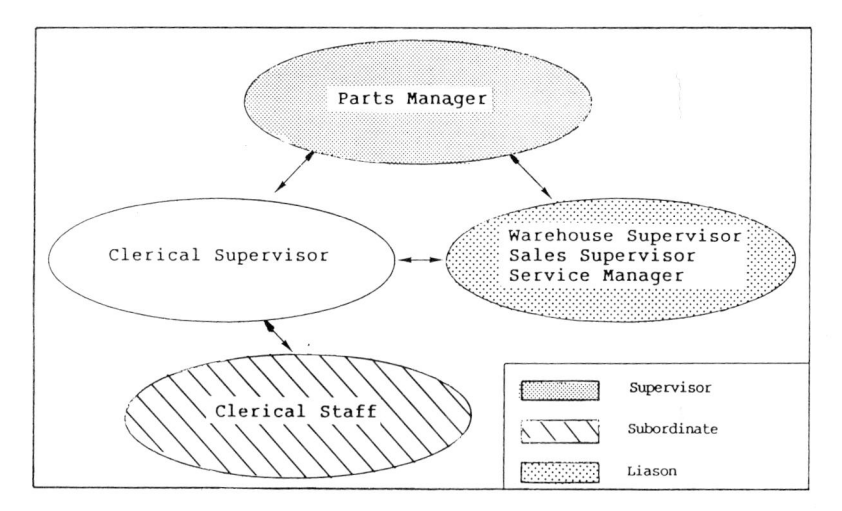

```
                    ┌─────────────────┐
                    │  Parts Manager  │
                    └─────────────────┘
                         ↙        ↘
   ┌───────────────────┐  ←→  ┌───────────────────────┐
   │ Clerical Supervisor│      │ Warehouse Supervisor  │
   │                   │      │ Sales Supervisor      │
   └───────────────────┘      │ Service Manager       │
             ↓                └───────────────────────┘
   ┌───────────────────┐
   │   Clerical Staff  │
   └───────────────────┘
```

▬ Supervisor
▧ Subordinate
▒ Liason

<u>Reports to:</u> The parts manager.

<u>Summary of responsibilities:</u> The clerical supervisor is responsible for the efficient operation of all clerical functions concerned with order preparation, accounting, stock records and general office procedures including typing and filing.

<u>Detailed job description:</u> The clerical supervisor has day-to-day responsibility for the following tasks:
supervision of entry on the stock record cards of
- customer invoices
- workshop requisitions
- supplier invoices
- orders
- prices and ancillary information;

supervision of regular reviews of the stock cards and calculation of fresh order levels and prices; supervision of order compilation;

correspondence with suppliers relating to orders;

supervision of back-order reconciliation with suppliers;

supervision of accounts clerks with book-keeping functions;

supervision of stock record clerks in entering price and availability information on customer orders;

supervision of pricing and recording for local order procedures;

carrying out staff training programmes on a day-to-day basis in accordance with the established programmes;

preparation of staff review reports.

Authority: The clerical supervisor has authority over the day-to-day activities of the clerical staff. The clerical supervisor has authority to make petty cash disbursements up to $...... on his or her own signature, using the standard procedures for petty cash accounting.

Liaison: The clerical supervisor liaises with the warehouse and sales supervisors on the day-to-day exchange of information necessary to operation of the department, and with other administrative departments to promote smooth operations and information exchange.

Standards: The clerical supervisor will have achieved a satisfactory standard when:

all stock record card entries are completed according to an established timetable;

all correspondence is kept up to date;

all incoming information from manufacturers relating to parts is entered on the stock cards within the same week;

order compilation timetables are properly maintained and orders are prepared and dispatched to suppliers on schedule;

book-keeping functions are performed accurately and kept up to date;

filing systems are properly maintained;

petty cash records are properly maintained;

management reports are prepared according to the agreed timetable;

clerical department staff maintain progress through pre-determined training programmes.

Clerical supervisor: Profile

Formal qualifications: The clerical supervisor must be educated to a reasonable standard of literacy and numeracy.

Techical training: The clerical supervisor does not normally need formal technical qualifications.

Experience: The clerical supervisor should have at least three years experience in a clerical position and should also have at least some managerial experience.

General character: The clerical supervisor should be extremely orderly and disciplined.

4.6 Sales supervisor

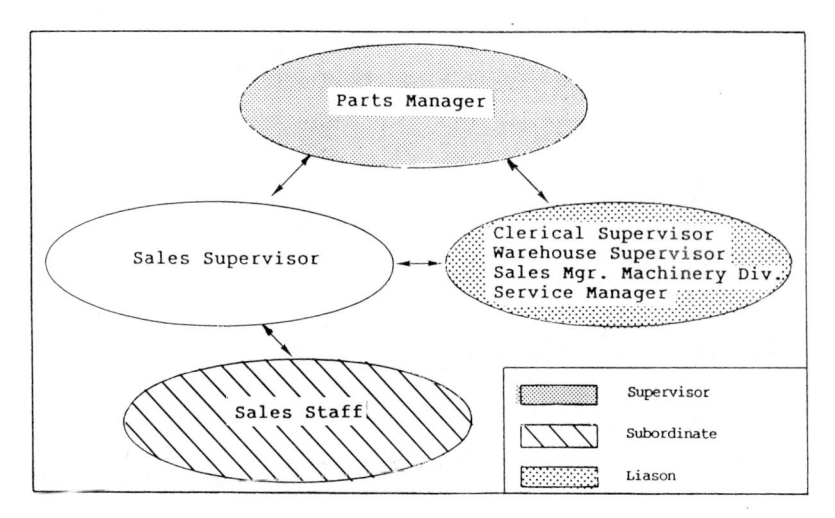

```
                    ┌─────────────────┐
                    │  Parts Manager  │
                    └─────────────────┘
                       ↑           ↑
                       ↓           ↓
  ┌──────────────────┐     ┌─────────────────────────┐
  │ Sales Supervisor │ ←―→ │ Clerical Supervisor     │
  │                  │     │ Warehouse Supervisor    │
  └──────────────────┘     │ Sales Mgr. Machinery Div│
           ↑               │ Service Manager         │
           ↓               └─────────────────────────┘
  ┌──────────────────┐     ┌───┐
  │   Sales Staff    │     │▓▓▓│ Supervisor
  └──────────────────┘     ├───┤
                           │///│ Subordinate
                           ├───┤
                           │:::│ Liason
                           └───┘
```

Reports to: The parts manager.

Summary of responsibilities: The sales supervisor is responsible for the maintenance of a working relationship between the parts department and its customers, and for ensuring that customer demands are recorded in such a way that the department can maximize customer satisfaction.

Detailed job description: The sales supervisor has the day-to-day responsibility for the following functions:

   reception of customers and the translation of their requirements into accurate customer orders;

   supervision of the correct identification of required parts;

   supervision of the task of finding substitutes for unavailable parts;

   supervision of back orders;

resolution of customer complaints;

supervision of invoice preparation;

supervision of cash receipts;

supervision of credit control;

maintenance of parts information, catalogues and microfiche;

orderly maintenance of counter areas, fixtures and fittings;

scheduling appointments for outside sales staff; implementation of staff training programmes on a day-to-day basis according to an established schedule;

design of promotional schemes to stimulate parts sales and implementation of these schemes once they are approved by management;

preparing staff review reports.

Authority: The sales supervisor has authority over the day-to-day activities of the sales staff. The sales supervisor has authority to make petty cash disbursements for the local purchase of minor parts needed for customer orders, providing such disbursements are properly recorded and substantiated according to standard procedures. The sales supervisor will have authority to offer credit to customers within the discount limits and guidelines established by the financial director.

Liaison: The sales supervisor works with the clerical and warehouse supervisors to ensure the smooth exchange of information necessary to operate the department. The sales supervisor liaises with the sales manager of the machinery division to ensure that the parts department has up-to-date information on machinery population changes, and with the service manager to plan maintenance and overhaul programme parts requirements.

Standards: The sales supervisor will have achieved a satisfactory standard when:

customers are satisfied in an efficient, timely and accurate manner;

customer complaints are resolved promptly;

customer orders are invoiced accurately and fulfilled immediately;

cash receipts are handled accurately and properly accounted for;

cash disbursements are immediately and properly accounted for;

credit accounts are kept up-to-date and overdue payments are collected regularly;

credit limits are observed;

sales targets are achieved;

customer service levels are maintained;

discount structures are properly observed;

back order records are properly maintained, regularly reviewed and effectively processed;

the sales counter, surrounding areas fixtures and fittings are maintained in a clean and tidy state;

parts catalogues and counter microfiche are up-to-date;

external sales appointments are properly reported;

sales staff reviews indicate that development is proceeding along planned lines.

Sales supervisor: Profile

Formal qualifications: The sales supervisor should have at least a high school-level education.

Technical training: The sales supervisor should have training in the technology of machinery maintenance and repair sufficient to provide a sound foundation for advising customers on the selection of parts for their equipment.

Experience: The sales supervisor should have at least three years experience in the supervision of sales staff in a mechanical environment. It is beneficial if previous experience includes the type of equipment handled by the organization.

General character: The sales supervisor should be an outgoing person who interacts well with others and can develop a reputation for fairness and honesty.

4.7 Warehouse chargehand

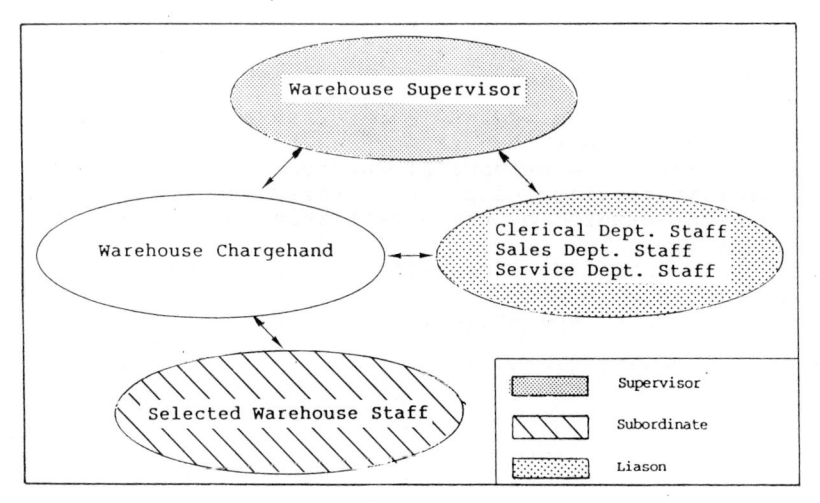

```
           Warehouse Supervisor

                                    Clerical Dept. Staff
   Warehouse Chargehand    ◄──►     Sales Dept. Staff
                                    Service Dept. Staff

      Selected Warehouse Staff      ▓  Supervisor
                                    ▨  Subordinate
                                    ░  Liason
```

Reports to: The warehouse supervisor.

General summary of responsibilities: The warehouse chargehand is responsible for the day-to-day supervision of the warehouse staff in carrying out a specific selection of tasks.

Detailed job description: The warehouse chargehand appointed to a specific section of the warehouse will take charge of the following:

   receiving documents relating to that section, for example, customer orders or supplier invoices;

   allocating specific tasks to warehouse staff;

   ensuring that the tasks are properly and efficiently completed;

   checking results in terms of location, identity and quantity of parts;

   reporting discrepancies or damage;

completing documents relating to the receipt or issue of parts;

ensuring store cleanliness and neatness;

carrying out day-to-day training of subordinate staff.

Authority: The warehouse chargehands have day-to-day authority over the warehouse staff allocated to his specific section. The senior warehouse chargehand may be expected to assume authority for the entire warehouse in the absence of the warehouse supervisor.

Liaison: The warehouse chargehands work with clerical and sales department staff to ensure the smooth operation of the warehouse, and with service department staff on questions of identity and issue of parts to the workshop.

Standards: The warehouse chargehands will have achieved a satisfactory standard of work when all tasks allocated to the warehouse staff are completed in a timely, accurate and efficient manner, all documents relating to the reception or issue of parts are completed within the time span allowed, and the warehouse is clean and tidy.

Warehouse chargehand: Profile

Formal qualifications: The warehouse chargehands should be able to read and write, but no formal education is required as long as it is understood that lack of education will restrict promotion.

Technical training: The warehouse chargehands need some technical knowledge of mechanical repairs for recognition of parts.

Experience: The warehouse chargehands should have some prior experience in a storekeeping environment.

General character: The warehouse chargehands should be capable of supervising subordinate staff and should be able to work in a conscientious, neat and accurate manner.

4.8 Warehouse staff

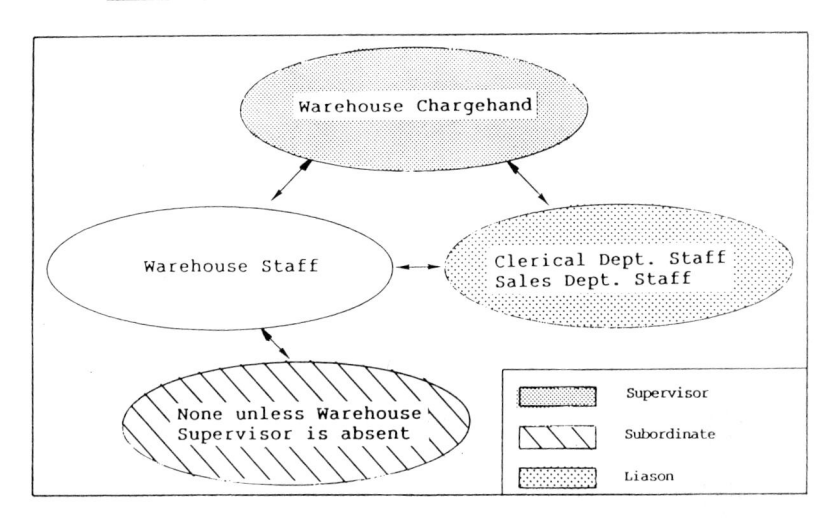

```
                    ┌─────────────────────┐
                    │ Warehouse Chargehand│
                    └─────────────────────┘
                         ↑         ↑
                         │         │
          ┌──────────────┐   ┌──────────────────┐
          │Warehouse Staff│◄─►│Clerical Dept. Staff│
          │              │   │Sales Dept. Staff  │
          └──────────────┘   └──────────────────┘
                 ↑
                 │
          ┌──────────────┐        ▓▓▓  Supervisor
          │None unless Warehouse│  ╱╲╱  Subordinate
          │Supervisor is absent │  ░░░  Liason
          └──────────────┘
```

Report to: The warehouse chargehands.

Summary of responsibilities: Warehouse staff are responsible for efficient execution of warehouse tasks delegated to them by the warehouse chargehands.

Detailed job description: The warehouse staff carry out the following tasks as required:

  receiving, unpacking and checking incoming parts;

  stocking incoming parts;

  passing picking lists and supplier invoices to the clerical department;

  recording any discrepancies or damage of incoming parts;

  recording location changes;

  picking parts in accordance with orders received from the clerical department;

assembling orders and checking parts;
delivering orders to the sales counters;

packing orders for outside delivery;

cleaning parts, shelves and the store generally, and removing rubbish;

reporting any irregularities such as shortages, errors or damage.

Authority: Authority is not delegated to warehouse staff except when the warehouse supervisor is temporarily absent.

Liaison: The warehouse staff liaise with the clerical and sales department staff to ensure smooth operation of the store.

Standards: The warehouse staff will have achieved a satisfactory standard of performance when the tasks delegated to them are completed in a timely, accurate and conscientious manner, and when the warehouse and its contents are kept clean and tidy.

Warehouse staff: Profile

Formal qualifications: The warehouse staff should be able to read and write, but no formal education is required as long as it is understood that lack of education will restrict promotion.

Technical training: The warehouse staff need some technical knowledge of mechanical repairs for recognition of parts. This can be learned from on-the-job training programmes.

Experience: The warehouse staff will benefit from prior experience in a storekeeping environment, but this is not essential.

General character: A warehouse staff member should be capable of following instructions and working conscientiously, accurately and neatly.

4.9 Records clerk

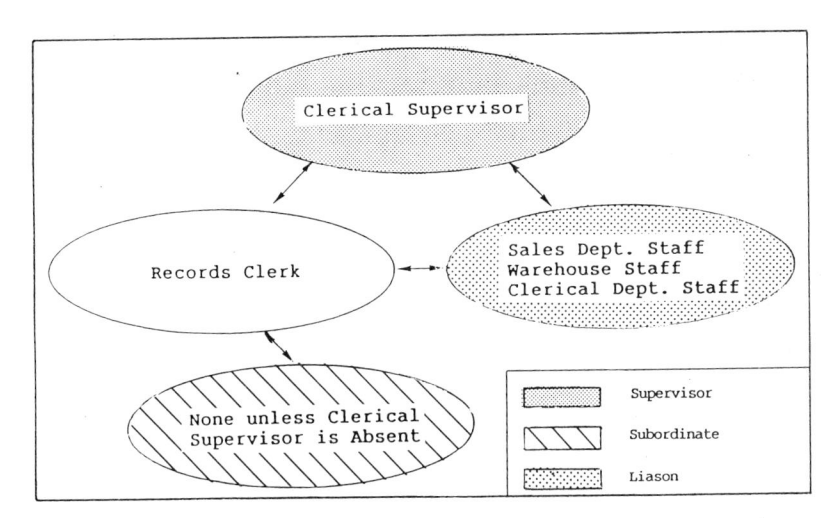

```
Clerical Supervisor
        ↓↑
Records Clerk ←→ Sales Dept. Staff
                 Warehouse Staff
                 Clerical Dept. Staff
        ↓
None unless Clerical
Supervisor is Absent

■ Supervisor
/// Subordinate
▨ Liason
```

Reports to: Clerical supervisor.

General summary of responsibilities: The records clerk is responsible for accurately maintaining stock record cards, and for using information from the cards to process customer orders.

Detailed job description: The records clerk enters the following information on the cards:

details of parts received according to supplier invoices received from the warehouse;

cost and price information from costed invoices;

parts number changes and cross references from suppliers' information bulletins;

details of parts issued gleaned from customer invoices;

stock balances;

marks to facilitate the order compilation process.

The records clerk will retrieve information from the cards to:

enter parts availability information on customer orders;

enter prices on customer orders;

enter parts number changes and on customer orders.

Liaison: The records clerk works with the sales department and warehouse staff to clarify details of orders and incoming parts. Cooperation with other members of the clerical department is necessary to ensure that access to the cards is shared by all who need it.

Authority: The records clerk has authority to make whatever changes are required on the cards to comply with information received; to create new cards where necessary for new parts numbers; to replace cards which are full; and to calculate prices in accordance with established rules. If the records clerk is the senior clerk in the department, he or she may assume authority for running the clerical department in the absence of the clerical supervisor. In this case, the clerical supervisor's authority for petty cash disbursement will be taken oven by the parts manager.

Standards: The records clerk will have achieved a satisfactory standard of work when all card entries are accurately posted within the allowed time span, and when customer order entries are completed promptly.

Record clerk: Profile

Formal qualifications: Record clerks should have a formal high school level education in literacy and numeracy.

Technical training: Record clerks will benefit from prior training in inventory control and office procedures, but this is not essential where it can be given within the organization.

Experience: They should have some prior experience of office routines, but this is not essential where they can be hired for junior positions and trained within the organization.

General character: Record clerks should be conscientious, neat and accurate.

4.10 Order clerk

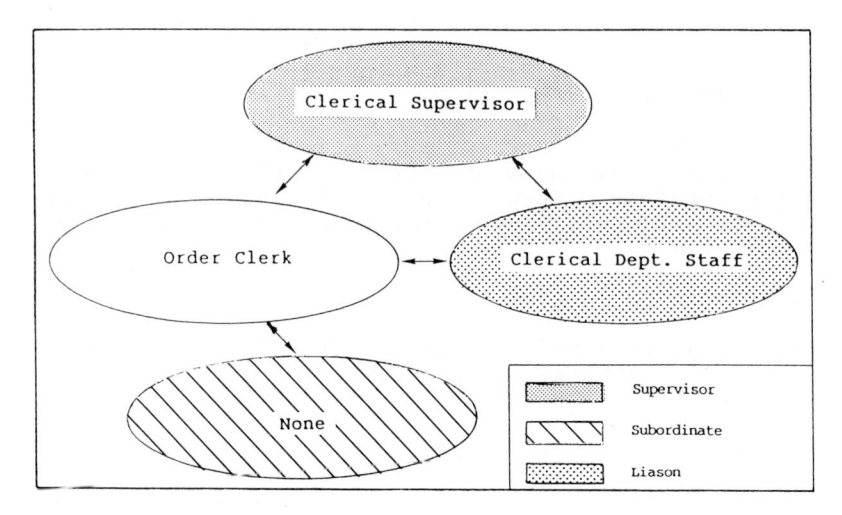

```
         Clerical Supervisor
          ↙            ↘
  Order Clerk  ←→  Clerical Dept. Staff
       ↓
      None

  ▓ Supervisor
  ▨ Subordinate
  ░ Liason
```

Reports to: The clerical supervisor.

General summary of responsibilities: The order clerk is responsible for compiling orders to ensure that stocks are adequately replenished.

Detailed job description: The order clerk reviews the stock record cards in order to:

identify emergency order requirements daily from cards showing appropriate marks, and compile orders for suppliers;

identify stock order requirements from cards with appropriate marks, and compile stock order indents for pricing and submission to management;

total the issues on a quarterly basis and recalculate order points and standard orders;

enter details of orders submitted to suppliers;

check regularly on supplier back orders and maintain communication with suppliers on back orders.

Authority: The order clerk has authority to compile orders in accordance with established procedures, and to correspond with the suppliers on the progress of such orders.

Liaison: The order clerk works with other members of the clerical department to ensure smooth operation.

Standards: The order clerk will have achieved a satisfactory standard of work when all cards bear regularly updated order reference information, when order compilation schedules are maintained, and when up-to-date information is recorded on all back orders.

Order clerk: Profile

Formal qualifications: Order clerks must have a formal high school level education in literacy and numeracy.

Technical training: Order clerks will benefit from prior training in inventory control and office procedures, but this is not essential where it can be given within the organization.

Experience: They should have some prior experience of office routines, but this is not essential where they can be hired for junior positions and trained within the organization.

General Character: Order clerks should be conscientious, neat and accurate.

4.11 Accounts clerk (for organizations where accounting is not centralized).

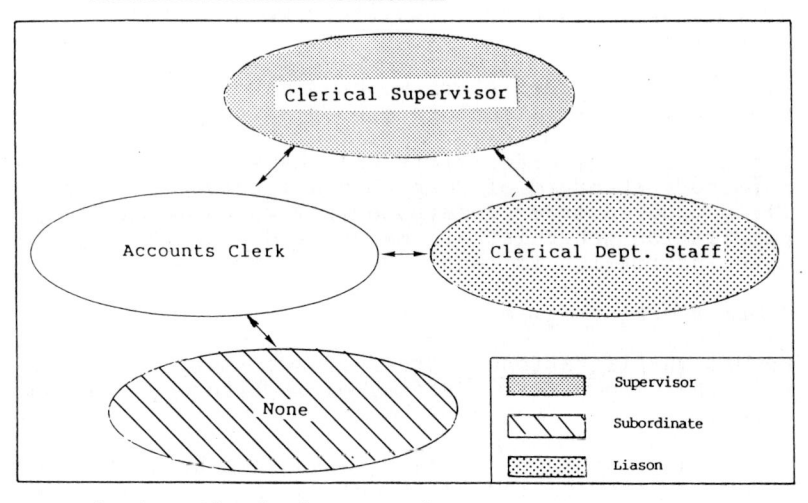

```
                    ┌─────────────────────┐
                    │ Clerical Supervisor │
                    └─────────────────────┘
                           ↑
                           │
    ┌───────────────┐      │      ┌─────────────────────┐
    │ Accounts Clerk│ ←──→ │      │ Clerical Dept. Staff│
    └───────────────┘      │      └─────────────────────┘
                           ↓
                    ┌─────────────────────┐
                    │        None         │
                    └─────────────────────┘

                    ▓▓▓ Supervisor
                    ╲╲╲ Subordinate
                    ░░░ Liason
```

Reports to: Clerical supervisor.

General summary of responsibilities: The accounts clerk keeps the department's financial records up to date, and prepares management reports based on these records.

Detailed job description: The accounts clerk enters the following in the appropriate accounting books:
  details of supplier invoices;

  details of ancillary costs related to parts supply;

  details of all expenditures related to departmental operation;

  details of all invoices, requisitions or other documents relating to the issue of parts.

The accounts clerk maintains records of credit accounts and brings to the clerical supervisor's attention all overdue accounts on a regular basis. The accounts clerk assembles cost of supply information to maintain accurate and up-to-date costing and pricing for parts. The accounts clerk gathers information from the records for the monthly preparation of management reports.

Authority: The accounts clerk has authority to set prices in accordance with established procedures. The accounts clerk has no authority for decisions or activities relating to cash handling or expenditures.

Liaison: The accounts clerk works with other members of the department to ensure smooth operation, and with other accounting staff to ensure uniformity of practices.

Standards: The accounts clerk will have achieved a satisfactory standard of work when all book-keeping entries are accurate and up-to-date; when no documents await posting to the books; and when management reports are presented accurately and completely, by the required date each month.

Accounts clerk: Profile

Formal qualifications: Accounts clerks must have a formal high school level education in literacy and numeracy.

Technical training: Accounts clerks will benefit from prior training in inventory control and office procedures, but this is not essential where it can be given within the organization.

Experience: They should have some prior experience of office routines, but this is not essential where they can be hired for junior positions and trained within the organization.

General character: Accounts clerks should be conscientious, neat and accurate.

4.12 Counter clerk (sales department: retail or workshop)

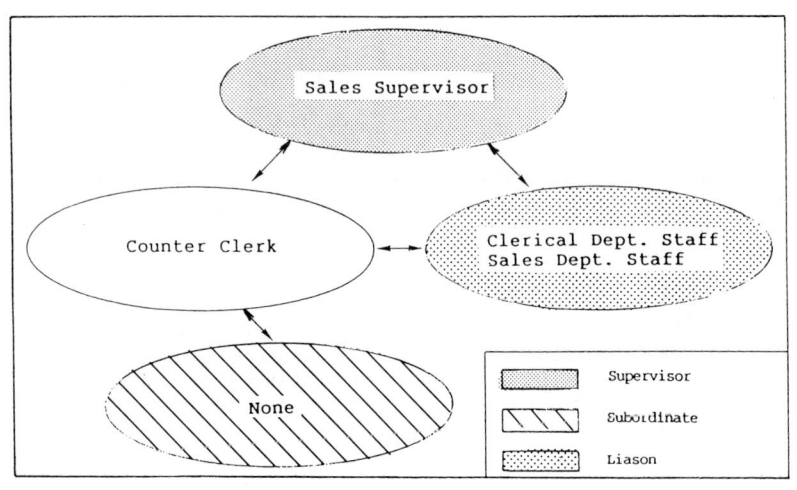

[Organization chart showing Counter Clerk reporting to Sales Supervisor, with liaison to Clerical Dept. Staff / Sales Dept. Staff, and no subordinates. Legend: Supervisor, Subordinate, Liason.]

Reports to: Sales supervisor.

General summary of responsibilities: The counter clerk is responsible for receiving all customer enquiries and for interpreting them to enable parts to be supplied to the customer's satisfaction.

Detailed job description: The counter clerk, on receiving an enquiry for parts, will:

write down the enquiry on a customer order form with details of the type of part, and the machine involved;

identify the part by its current part number from the latest catalogue or microfiche;

pass the customer order to the clerical department for availability and price entries, and for picking by the warehouse;

receive the parts and the completed customer order from the warehouse;

check the parts with the customer, identify any changes and explain them;

discuss any back orders and obtain the customer's confirmation that back-ordered parts are still required;

complete any further details on the invoice, and calculate the invoice total together with any taxes, ancillary charges and/or discounts to be applied;

collect cash for the invoice, or hand the invoice to the cashier who will collect the cash, or check the customer's credit and if this is in order, obtain the customer's signature on the invoice and delivery slip;

perform the same tasks on the invoice of parts ordered by mail to be delivered to the customer;

regularly review all customer open order files and check the progress of back orders;

invoice back orders as they arrive;

contact customers verbally or in writing on the progress of their back orders;

identify customer requirements for non-stock items for local purchase;

maintain lost sales records for parts not supplied; keep the counter clean and neat;

resolve problems with customers;

maintain cordial relations with all customers or potential customers to promote the organization's business and maximize sales;

discuss parts requirements with customers and make suggestions about additional needs customers may not have identified.

Authority: The counter clerk has authority to:

write invoices in accordance with agreed procedures relating to prices and discounts;

collect cash for invoices (unless a cashier does this);

accept parts returned by customers according to agreed procedures;

accept customer orders for future supply including those for back-ordered parts;

make substitutions where they result in better customer satisfaction.

The counter clerk does not have authority to reduce or increase prices except within agreed discounting practices. The counter clerk does not have authority to offer credit except within the rules and limits established for individual customers by the financial controller.

Liaison: The counter clerk works with other departmental staff for smooth operation, and with other sales staff in the organization in the interests of customer satisfaction.

Standards: The counter clerk will have achieved a satisfactory standard of work when:

customer orders are dealt with promptly and accurately;

invoices are completed accurately;

disputes over parts supply or invoicing are handled competently and to the satisfaction of both the store and the customer;

discount and credit procedures are handled correctly;

back orders are regularly processed;

lost sales reports are maintained;

high customer service levels are maintained; no customers are lost to competitive suppliers through lack of attentiveness by counter staff.

Counter clerk: Profile

Formal qualifications: The counter clerk should have enough technical knowledge to recognize parts and identify customers' requirements. This need not have been with the specific makes and models of equipment to be handled.

Experience: The counter clerk should have some prior experience dealing with customers in a storekeeping or workshop environment.

General character: The counter clerk should be attentive and helpful, capable of resolving customers' problems and attentive to their needs in a patient and amiable manner.

4.13 Other staff

Job descriptions similar to those above can be compiled for other members of the parts department staff such as typists, filing clerks and drivers, whose tasks are less specialized, but still necessary.

None of the above job descriptions refer to stocktaking. In organizations where this is an annual event, it is often necessary to use all the staff to count parts and record items and quantities. Stocktaking is organized by the parts manager under the supervision of the financial controller or auditor.

In organizations using a perpetual inventory system, a specific person, aided perhaps by a team of labourers and clerks, has the responsibility for stock-checking on a full-time basis. The job description for this post is set out below.

4.14 Perpetual inventory supervisor

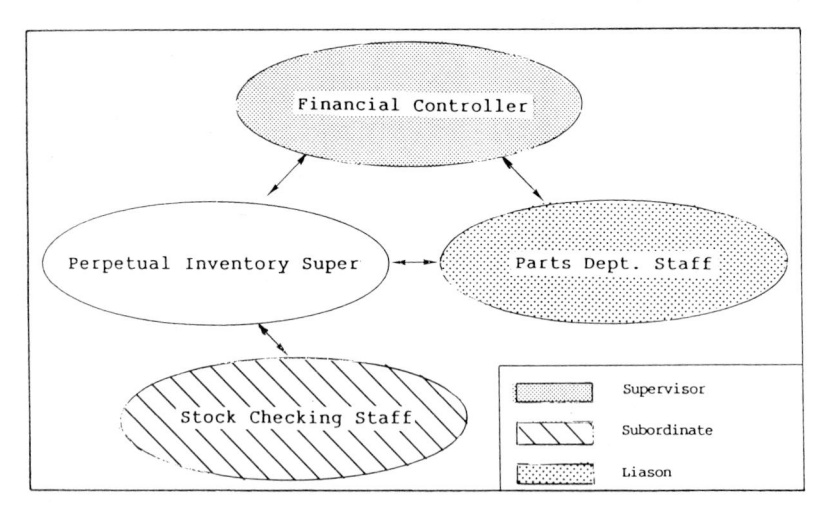

[Diagram: Financial Controller (Supervisor) linked to Perpetual Inventory Super and Parts Dept. Staff (Liason between them); Stock Checking Staff (Subordinate) linked to Perpetual Inventory Super. Legend: Supervisor, Subordinate, Liason.]

Reports to: Financial controller (or internal auditor).

General summary of responsibilities: The perpetual inventory supervisor is responsible for the physical check of stocks against the theoretical stock quantities shown in the stock records, and for reporting and investigating the cause of all discrepancies.

Detailed job description: The perpetual inventory supervisor establishes, for the purpose of planning work programmes, movement rates for the various types of stock in the warehouse. The perpetual inventory supervisor makes random checks on individual bins or entire shelf units in such a way that the physical count is achieved rapidly and without impeding the normal operation of the department. The actual count is compared immediately with the stock balance figure on the record cards. If the two figures agree, the card is marked to show that the balance is correct. If the figures do not agree, the perpetual inventory supervisor then makes a series of checks for:

- arithmetical errors on the card;

- invoices not posted;

- parts misplaced in other bins;

- incorrect identification of similar parts;

- short shipments not properly identified;

- any other possible causes peculiar to the organization.

Whatever the cause of the discrepancy, the perpetual inventory supervisor records the correct stock figure on the card and makes a report to the financial controller.

Authority: The perpetual inventory supervisor has authority over the daily activities of any staff assigned to the stock checking team. The perpetual inventory supervisor has the authority to stop the delivery of parts to or from the sections of the warehouse under investigation until checking is complete.

The perpetual inventory supervisor has authority to correct discrepancies on the stock cards.

Liaison: The perpetual inventory supervisor works with the parts department staff so that their work is not un-necessarily impeded by the stock-checking process.

Standards: The perpetual inventory supervisor will have achieved a satisfactory standard of work when, at the end of the year, all stock items have been checked the requisite number of times, all discrepancies have been investigated, and those not subject to simple correction have been reported in such a way that the auditors are satisfied.

Perpetual inventory supervisor: Profile

Formal qualifications: The perpetual inventory supervisor should be educated to a reasonable standard of literacy and numeracy.

Technical training: The perpetual inventory supervisor should be well trained in parts identification.

Experience: The perpetual inventory supervisor must have at least three years clerical experience with a machinery maintenance and repair organization.

General character: The perpetual inventory supervisor should be extremely orderly and scrupulously honest.

4.15 Staff selection

Hiring is not easy; it demands an assessment of personal capabilities and attitudes. In many countries, personnel selection has become a service industry in itself.

Finding the right staff depends on selecting from as large a number of applicants as possible. This is easier for positions requiring fewer skills. It is much more difficult in the case of more senior staff who must have the specialized knowledge and experience vital to the enterprise.

New staff may be recruited through: government employment offices; private staffing agencies; a prominently-displayed notice; word-of-mouth among present staff and their acquaintances; press advertisements. The method chosen will vary with the post to be filled. For example, for the more sensitive senior or highly-paid posts, it may be necessary to search among rival organizations for a person of proven capacity. On the other hand, labourers may be recruited by posting a notice on the gate.

An applicant's employment record, or curriculum vitae, should show steady progress in the acquisition of knowledge and experience to reach the point at which the applicant has the qualifications for the post offered. An employmernt record should also be checked for any unexplained gaps in employment which might indicate forced or voluntary unemployment. These may reflect the general unemployment situation of the area or indicate a health problem.

To ensure that all essential information is obtained, an applicant should be asked to complete an employment application form as the initial step in the selection process. When there are many applicants for a post, primary selection can often be based on the application form and the majority of unsuitable candidates eliminated. A sample application form is shown on page 108.

Personnel Application Form

Company name
Address
Vacancy _____ Ref. _____
Application from

Family name _____ Other names _____
Address   Telephone no. _____
                          Married ____ Single ____
                          Children (ages) _____
Date of birth _____ Age ____ Nationality ____

Degrees and/or technical qualifications
Driving Licence? ____ Criminal convictions? ____
Schools attended (give dates and certificates)
Present position _____ From ____ Salary ____ Name of employer _____ Address   Telephone no. _____
Brief details of duties
Name of supervisor who may be contacted for reference
Please indicate if you do not wish present employer to be contacted. _____
Signed _____ Date _____

Employment Record
(Reverse of Personnel Application Form)

| Employer _____ From _____ To _____ |
| Address   Telephone _____ |
| |
| Position_____ Salary _____ |
| |
| Reason for leaving |

| Employer _____ From _____ To _____ |
| Address   Telephone _____ |
| |
| Position_____ Salary _____ |
| |
| Reason for leaving |

| Employer _____ From _____ To _____ |
| Address   Telephone _____ |
| |
| Position_____ Salary _____ |
| |
| Reason for leaving |

Some organizations contact applicants by telephone. If the person interviewing by phone has a structured list of questions, unsuitable candidates can be eliminated at that stage. After the initial elimination process, a "short list" of those candidates who meet the requirements can be compiled and interviews arranged.

If the organization has a personnel department, the selection process is usually organized there. Nevertheless, the manager of the department employing the candidate should conduct the interview.

During the interview, the candidate should have the opportunity to ask questions about the job and the organization. However, the interview should be conducted formally with properly-planned questions and recorded answers. For a recorded interview, a written format may be used. Questions should be laid out in a logical sequence and designed to ensure that a complete record is established of the applicant's knowledge and experience. An example of a suitable format is provided on page 111.

References from previous employers and respected members of the applicant's community should be interpreted with care as they may be influenced by personal relationships. Some personnel specialists speak directly to references by telephone.

4.16 Training

Training should be a continuous process within the organization. It will seldom be possible to find staff with the exact qualifications needed to fill every position. An employee's previous employers may have used different methods or handled other types of equipment. Therefore, any staff taken on will have to be trained in the precise policies and procedures they will be expected to use.

There will also be some normal staff turnover. As the organization expands, employees may be promoted or moved to fill new vacancies, or staff may leave.

Structured Interview

Interview by _____ Date _____   By telephone? ____ In person? _____   Name of reference _____ Position _____   Address   Telephone _____
Name of applicant _____   Present position _____ From _____   Duties   Competence   Reliability   Timekeeping Good? ____ Poor? ____   Absenteeism Yes? ____   No? ____
Qualifications (check applicant information)   Education (check applicant information)   Employment record (Check application information)
Recommendation based on this interview   Signed _____

This means an intermittent need to teach how to perform new functions. Overlaying this intermittent process, however, is a general need to develop employees' capabilities; to involve them in a continuous process of education so that the functions of the organization are continually improving. This also contributes to motivation as employees realize that training opens up new pathways to promotion and personal achievement. Also, the equipment serviced and its constituent parts change constantly. Manufacturers introduce new models and modify old ones, calling for a regular training programme to keep staff up to date on technical developments.

It is sometimes suggested that training is counterproductive if it equips employees to seek better-paid employment elsewhere. Such fears can be countered in two ways. First, an organization with well-trained staff can usually function at an enhanced level of efficiency and profitability and maintain a competitive salary structure. Second, if employees view a training programme as a process by which they can develop and progress within the organization, they may be less inclined to leave for an unclear future elsewhere.

The training programme should be both general and specific, that is, it should have a syllabus designed to bring all staff members up to required standards of knowledge, whatever their previous experience or training. But it should also be flexible enough to be tailored to the specific needs of each staff member.

If the programme is designed as a combination of several courses on different topics, training sessions can be conducted for groups of staff from different departments. They will progress individually from topic to topic, not always in the same group, until each individual reaches a satisfactory level. Slow learners will repeat sections of the course while more rapid learners pass to the next section.

All members of staff should have some knowledge of the work of other departments. For example, it may be too time-consuming to train warehouse staff in advanced inventory control or invoicing, but it is important that they have an awareness of office procedures and even sales work. Training may reveal talents which eventually enable warehouse staff members to qualify for promotion to clerical or sales duties.

Everyone should be included in some aspect of the training programme. Senior management will make objective decisions about the training that is needed for the parts manager. Particular attention must be paid to making sure the parts manager is kept up to date on products. In addition, the manager should not ignore any opportunities for self-improvement in the technology of parts systems and administration.

Training programmes for the rest of the staff should be organized in a detailed, structured schedule; an example is shown in Appendix C. The programme should be based on a one-year cycle and a new programme should be set out before the old one is finished. Not only will this process be seen as a continuous one, but staff members will also see how it fulfills their needs. Individuals should be encouraged to ask for training in areas they feel would benefit them or where they believe they need a refresher course.

A structured training schedule ensures that the programme proceeds in an orderly and logical sequence in which elementary knowledge is implanted first, and each subsequent subject is built on a firm base. Material for training programmes may be drawn from many sources. Parts manufacturers supply regular information on modifications and parts changes.

They also supply information on maintenance and repair techniques, and usually have their own ideas about storage and inventory control technology. Management text books are useful for courses on administration techniques. Storage equipment manufacturers and the materials handling industry in general produce a great number of interesting manufacturers' catalogues and trade magazines. As the use of computers increases, there is a corresponding increase in publications related to both equipment and computer programmes.

While these sources can provide regular new training materials, the main content of the programme is normally the basic instructions which enable the staff to carry out the organization's policies. There may be important changes in these policies as new lines of equipment are added or old ones dropped. New executives in the organization may adopt different strategies. However, the basic day-to-day business of parts acquisition, storage and delivery, is controlled more by the unchanging laws of mathematics than by changes in management or equipment.

Although it is strongly suggested that training should be continuous, this does not mean that staff will spend long periods away from their normal jobs. In fact, one of the virtues of continuous training is that it can be carried out in short spells, at convenient times and in a way that does not disrupt the work of the department. It may be possible to get staff to attend training sessions after normal working hours on their own time, encouraged by a policy of basing promotion, bonus payments, or certain pay increases on completion of specific training.

Staff may expect overtime payment for any attendance after normal working hours, in which case, the training programme must be fit into the normal working day. Sessions should be kept short and should be timed to coincide with slack periods. In a large organization, each sub-division of the department may have enough staff to cover vital functions while others attend a training session. Smaller organizations will find this more difficult. The important point is that a regular programme should be set up, even if it is only one hour a month.

Some training may come from outside the organization, but the main tasks of training should be the responsibility of the organization's own managers and supervisors. In a large organization, a training department may be set up with a schoolroom, equipment and a full time staff. In any case, supervisors are responsible for training junior staff. In explaining a procedure to others, the trainer becomes more familiar with it; the supervisor thus becomes a better supervisor.

External training should also be used when possible. For instance, major manufacturers or suppliers normally make periodic sales and service visits to observe how their products are handled. If requested, the representative can prepare material and training aids in advance. Some manufacturers run their own training schools to which staff may be sent for training courses, or from which instructors may be brought for local training sessions.

Vocational training institutions may also provide useful courses. It is not unreasonable to expect ambitious staff members to attend evening classes voluntarily. In some cases, day release training may qualify for government support. This is especially important where new technology, such as a computerized administration system, is introduced.

Film projectors are necessary for some suppliers' instructional materials such as films or film strips. An overhead projector for diagrams or parts illustrations is also useful. Most administrative training requires only a blackboard or flip charts for adequate illustrations. A wide range of instructional material from machinery manufacturers and specialist companies is available for technical training on maintenance and repairs.

Basic courses for junior staff, particularly those for new staff, should be run as integral blocks. Other courses may be split up and scattered throughout the programme. The year's training programme should be broken down in a weekly calendar and posted prominently so that staff may remind themselves of the programme dates. Each staff member should be given an individual training programme. A copy of this, together with reports on the employee's progress, should be kept in the personnel files.

Training courses should be interactive, that is, staff should make as much of a contribution to the sessions as the instructor, either by answering questions, relating their own experience of a situation or by suggesting possible improvements or new ideas. A written test should be included in each course, the results of which provide a record of the success of the instructor as well as a record for the employee's personnel files. The latter is only a partial record however. The supervisor's daily and weekly reviews of staff performance give a better picture of the learning capability of employees.

4.17 Staff control

Staff control falls into two main categories: the daily maintenance of discipline and order; and the process of development.

Discipline is more easily maintained where the policies, objectives and rules of the organization are clear to all staff members. This is where a clearly-written "Terms and Conditions of Employment" document is useful, administered as a letter to each new employee, or as a printed booklet. In either case, the document should detail the organization's background, structure, policy and staff development programme, along with the rules and regulations for daily conduct. The document should also cover working hours, holidays, sick leave, health facilities, uniform allowances and any other employee benefits.

Rules should be specific and the results of ignoring them should be clearly spelled out. The tendency for laxity to grow with time can be combatted if the training programme syllabus includes a review of policies and rules on a regular basis.

The ultimate sanction, dismissal of an employee, is normally covered by local rules embodied in trades union agreements, in employment legislation, or both. The procedures may vary according to the seniority; whether the employee is paid monthly or weekly, still on probation, etc.

Discussion of employment legislation is outside the scope of this publication. However, a point worth emphasizing here is that rules about dismissal usually call for a certain number of verbal or written warnings. Often, managers and supervisors do not realize until it is too late that the first infringement of a rule must provoke a response, however mild. If habitual rule breaking makes dismissal a necessity, the proper series of warnings will have been put in place and the organization can defend itself against claims of unfair treatment.

When a trade union is involved, many terms and conditions of employment must be negotiated and will form part of the union contract. This makes the task of compiling an employee's information document somewhat easier since relevant portions of the union contract can be included word for word. Salary rates will also be part of the union contract.

When no trade union is involved, it is still advisable to establish formal pay scales which may or may not be revealed to employees, depending on the organization's policies. The principles of a salary structure are that people should be paid more or less the same rates for similar tasks and similar responsibilities; those with more responsibility merit higher pay. A margin of flexibility ensures that seniority, loyalty and an exemplary attitude can be rewarded. For this reason, a range of pay should be outlined for each post for a logical progression from one level to the next with little or no overlap. This ensures that senior employee's pay is not overtaken by that of junior staff. It also offers junior staff an incentive for advancement.

Employees may increase their pay by working overtime. This option may be limited to hourly, or in some cases, weekly paid employees; it is seldom offered to monthly paid, salaried staff who are assumed to have reached a level of seniority where working hours relate more to the completion of tasks rather than the clock. Whatever policy is adopted regarding overtime, it should be put in writing and subject to budgetary estimates and limitations. Excessive regular overtime may indicate the need to increase staff. Generally, overtime pay should not be more than 15-20 percent of an employee's wages, but it can increase to as much as 35 percent during periods of stocktaking.

Another way employees can increase their earnings is through bonuses, usually related to increased productivity. In the simplest case, sales staff may be paid partly on commission. Sales people are occasionally paid only on commission, but it is normally better to pay a living wage and augment it based on a percentage of sales or gross profit achieved.

The sales commission system may extend to counter staff in order to encourage customer satisfaction. However, this also requires a supporting effort on the part of the warehouse and clerical staff. It is often better to provide a productivity-related bonus to all staff including the manager. Most important in designing a bonus system is that the reward or bonus should be very clearly linked to the effort that produces it, and that the effect on wages should appear quickly - usually on the monthly management reports which provide the basis for bonus calculation.

The second major area of staff control relates to staff development. If the goals and needs of the organization are clearly set out in structural charts, detailed job descriptions and employment profiles, the capabilities and performance of the staff can be assessed against an established set of standards. The training programme can then be designed to produce a measurable improvement in this match of existing capabilities against theoretical standards. Control is required to ensure that this process is properly documented for each individual, and to ensure that the success of the programme is monitored so that weaknesses or omissions become apparent. The staff development programme is then useful for meaningful development.

V. THE STORE: FACILITY AND ORGANIZATION

The capacity required in a parts store depends mainly on the number and type of machines the store will serve. There are various ways to determine the quantity of parts that will be required. Over its lifetime, a tractor generally needs repairs and maintenance costing about the same as its original price. In other words, servicing costs are about the same as the true rate of depreciation. Some types of farm machines, such as disc harrows, require up to four times their original cost to keep them running under tropical soil conditions. Simple machines, such as field rollers, need much less - possibly only half the initial cost. These expenses are not evenly spread over the life of the machine; they may approach 25-30 percent of initial cost in years when the machine needs major overhauls.

Service costs include both parts and labour. In industrial countries, the high wages and readily available parts means that about 60 percent of repair costs may be labour charges. In a developing country where labour is cheaper and transportation costs increase the price of parts, the ratio may be 60 percent parts and 40 percent labour. These are rules of thumb and must be modified according to local circumstances.

Based on the above figures, a mixed tractor population will need 60 percent of its value in parts over an eight-year lifetime, or 7.5 percent of its value per year. In a fleet where all equipment is bought at the same time, there may be one or two years where parts needs reach 30 percent of the total value of the fleet.

Assuming that the average value of a medium horsepower wheeled tractor is US$11 000, then 7.5 percent represents a parts demand per tractor of US$850 if parts are bought at landed cost. The retail value of these parts would be about US$1 700.

What does this represent in terms of storage space? The normal single story parts store can support about 150 kg per square metre of overall storage floor space. If there is a second floor, the capacity will be 250 kg per square metre of ground floor space.

The cost of parts per kg varies greatly between different types of component. Small, intricate, electrical or fuel injection components cost much more than filters or castings. The cost of shipping parts by air can sometimes cost more than its FOB value. If a series of invoices represents all the suppliers involved, it can be analyzed to determine a rough parts-per-kg cost.

In 1982, the value of about 2.5 tonnes of truck parts shipped by air to Bamako, Mali, was US$13/kg - almost exactly the same value per kg as 1.4 tonnes of varied tractor and other parts shipped by air to Jamaica at about the same time. The sources were as far apart as Japan, Germany and England. If air freight charges were US$2/kg more than ocean freight, the same shipment sent by sea would cost US$11/kg.

If each tractor needs US$850 per year in parts at US$13/kg, this represents 65 kg of parts. Parts bought retail would cost US$26/kg. Where major capacity must be reserved for a fleet overhaul programme, some US$2 000 per tractor unit will be needed (30 percent of $11 000 x 0.6) in some years, representing 154 kg of parts.

Another simple calculation shows that for a medium horsepower tractor weighing 5 350 kg, replacement parts represent 1.2 percent of its weight. Figures published elsewhere indicate values between 0.6 percent (Fiat in a European marketing publication), and 5 percent (FAO Bulletin No.30 referring to a tractor working 2 000 hours per year).

The storage floor space needed, therefore, can be calculate as the number of machines to be serviced multiplied by a figure between 65 and 150 (depending on the population), and divided by 150 (or 250 for a two-story building). To this should be added the space required for offices and receiving and dispatching parts - probably 25 percent of the storage floor space.

A mixed fleet of 1 000 medium horsepower tractors will therefore need 540 sq m of floor space, (or 325 sq m in a two-story building). Tractor implements also require parts space which can be worked out using the service/depreciation factor mentioned above.

When planning a new enterprise, the possible necessity of future expansion should be kept in mind. One way to allow for that possibility is to start with a two-story building and construct a mezzanine floor when extra space is needed.

The turnover rate is another factor which affects space requirements. The examples above suggest that one year's parts must be stocked at all times, indicating a 1.0 turnover rate. In a commercial enterprise this would be uneconomical, and in an overseas operation, a rate between 1.0 and 2.0 would be the aim. If storage space is based on a turnover rate of 1.5, then only 360 sq m (540/1.5) would be needed.

5.1 Design and construction

Detailed building design and construction plans are outside the scope of this publication, but from a parts storage viewpoint, certain features must be considered.

The building should be weatherproof and reasonably dust-proof. There is a tendency in hot countries to design for maximum ventilation, but this should not mean unrestricted entry of dust; openings should at least be baffled. Louvred glass windows provide light and can shut out dust. Air conditioning may be needed, with the office area given priority and the cooled air then directed into the rest of the store.

Since electricity is expensive, maximum use should be made of natural lighting. Windows built high in the wall with external shading from a generous roof overhang provide indirect light and do not prevent the use of wall space for storage. However good the natural lighting, some artificial light will be needed for proper illumination of all storage areas. Fluorescent tube lighting is economical and does not create excessive heat. Light fixtures should be hung between the shelf units at a height which gives maximum illumination without impeding movement of goods stored on top of the shelf units. When planning the electrical installation, power supply sockets should be provided to facilitate the use of appliances such as vacuum cleaners and lead lamps.

An almost square building is better than a very rectangular one from the point of view of ease of operation and economical storage. Building height is likely to be influenced by factors such as adjacent workshops. If the parts store is in a section of a high roofed building, storage space will be better ventilated and it may be possible to add a mezzanine floor at a later date to increase storage capacity. When a high-roofed building is not available, ceiling height should be a minimum of 3 m.

There should be at least one double-width door to permit delivery of large cases of parts. This may be the only direct access to the store from outside. All other entrances can lead into the office or counter areas for security reasons.

The floor should be concrete, suitably reinforced to bear parts, the racks and shelf units. Load bearing capacity should also take into consideration the possibility of shelf height increases or the addition of a mezzanine. This means that while the overall bearing capacity for parts and fixtures should be a minimum of 500 kg per square meter (100 lbs/sq ft), point loads of up to 10 tonnes may be required under some circumstances. The floor should be smooth, well finished, and sealed to prevent it from becoming dusty. Painting the aisles and open work areas will make it easier to keep them clean.

Offices are necessary for the manager, clerical staff and stock control cards. Depending on the size of the operation, these may be subdivided offices. The manager should have a separate office with windows which provide a view of the other work areas and, if possible, of the warehouse.

The counter area should be separate from the offices with adequate space on both sides for counter clerks and customers. The counter length depends on the number of counter staff employed; it should allow about a meter of space for each to work in. Microfiche readers and any other counter furniture, such as catalogue racks, computer terminals or cash tills also require space. In the customer area, there may be seating, display cases, or, in some commercial stores, self-service units containing fast-moving items to stimulate impulse buying. A separate counter should be provided for the workshop.

Replacement Parts Store Layout

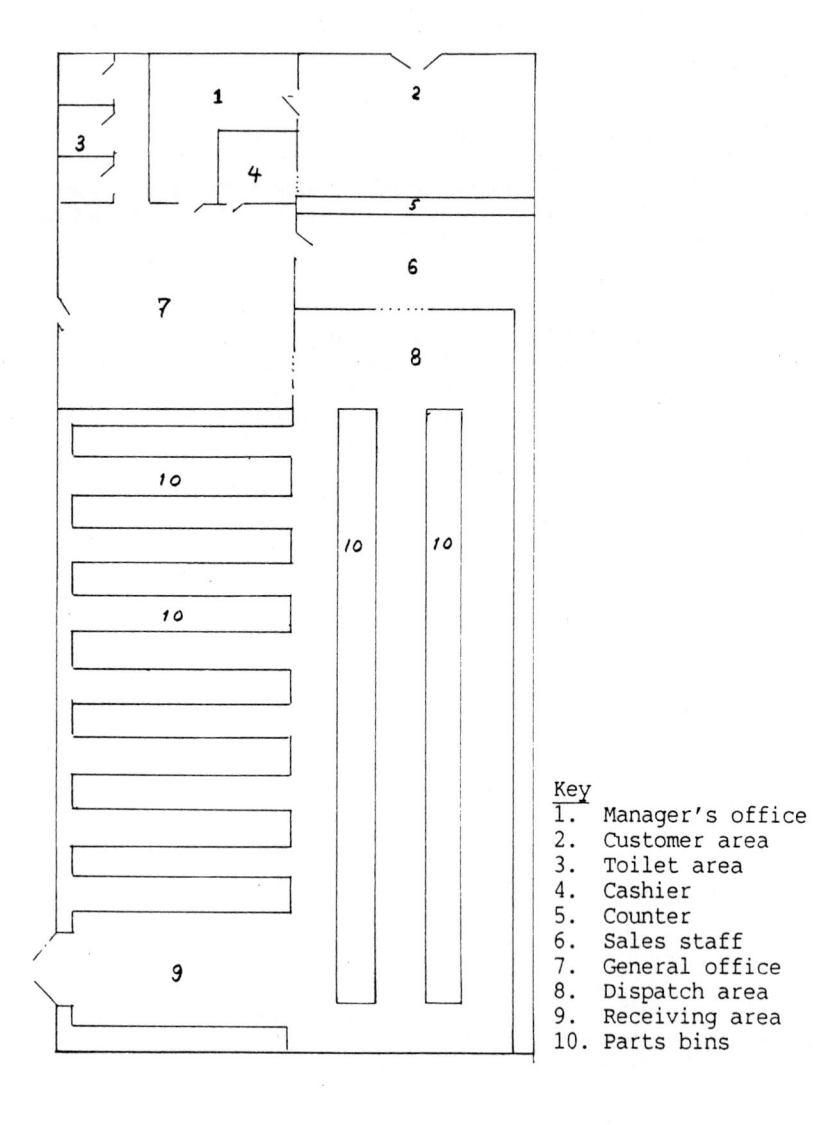

Key
1. Manager's office
2. Customer area
3. Toilet area
4. Cashier
5. Counter
6. Sales staff
7. General office
8. Dispatch area
9. Receiving area
10. Parts bins

5.2 Layout

The arrangement of the storage units within the building should provide good access to parts while making maximum use of available space. A very high degree of space utilization can be achieved with special stock picking equipment and the use of conveyors, but without these, 40-50 percent efficiency is a good target.

There should be adequate space between the shelf units for staff to arrange, clean and search among the parts. Cramped working conditions will lead to avoidance of housekeeping chores and errors in location and picking.

Aisles should be at least 80 cm wide where easily identifiable parts with a slow to medium rate of movement are stocked. Smaller, less easily identifiable parts with a faster rate of movement require aisles of about 90 cm. The main corridor through the store should be at least 1.25 m wide.

Access routes to larger parts need to be wider than suggested above. If mechanical handling equipment such as a forklift is used, space will be needed for turning and stacking. Equipment manufacturers can advise on the requirements for the various models.

In a large warehouse, fast moving stock may be divided into two portions. A quantity sufficient for a few days may be stored on a shelf near the counter while the bulk stock is kept on pallet racking in a different area. The day-to-day stock will be handled manually, while bulk quantities are moved mechanically.

Some parts are not suitable for shelf or bin storage. For example, steel bars, drive shafts and vehicle leaf springs are better stored on racks because of their length. Rack storage may be concentrated in a specific part of the store.

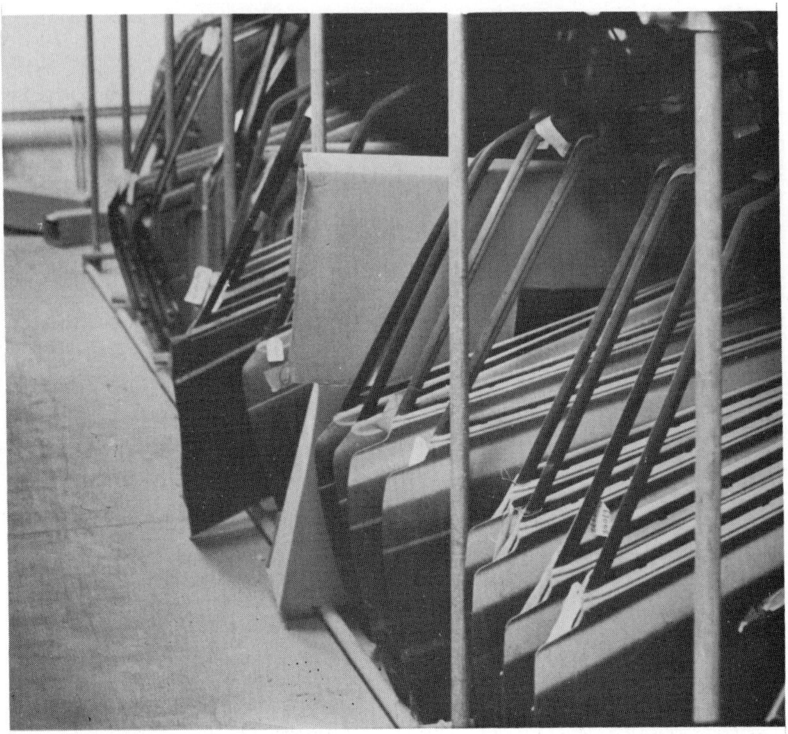

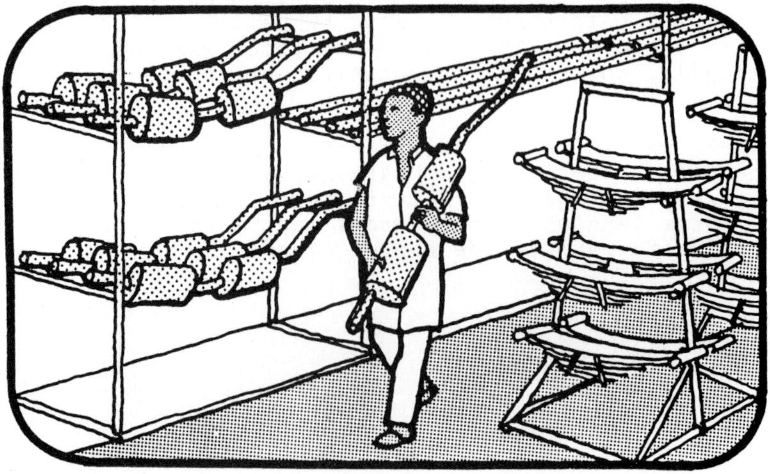

Large, relatively light items such as body panels are difficult to fit into shelves and may be placed either in special racks or on top of the shelving units where they may even span the aisles.

Wall space is wasted unless it is fitted with racks or shelves. Shelves against walls however, should be limited to 300 mm in depth since access is only available from one side.

Parts should not be stored on the floor. However, open floor storage may be needed for large items such as complete engines which remain boxed and can be stacked. In this case, the storage area should be marked with lines painted on the floor or there will be a tendency to stack goods in the aisles.

Two, clearly-marked open areas should be included in the layout design: one for goods received where incoming parts are unpacked and checked against invoices and packing lists before they are moved onto the shelves; and a second area for assembling and packing outgoing orders. These two areas should be well separated. If most parts go out over the counter, the receiving area may even be at the opposite end of the store from the dispatch area. On the other hand, if parts go in and out by truck, both receiving and dispatch areas may need to be adjacent to the truck dock.

In the loading and unloading areas of high-volume stores, different ground levels can sometimes be used to create a truck dock in which vehicles can park with their load decks at floor level. Otherwise, loading ramps or lifting devices must be considered.

Customer orders, invoices and picking documents can pass between the counter, the stock card index and the warehouse through hatchways in walls or doors. In a large store, documents can be sent by pneumatic document conveyors.

5.3 Shelving and storage equipment

Trends in shelf design have changed. At one time it was assumed that wooden shelf units were cheap and easily constructed from local materials. In some places this may still be true and, if so, wooden shelves can be robust and durable. The alternative to timber was the specialized steel parts bin system, complete with shelf dividers and special sections equipped with nests of little drawers. This is still available but tends to be expensive.

An alternative to building shelf units entirely of timber is to use angle iron for the end frames and shelf supports, and to make the shelves of wood. This offers the economical advantage of timber, while angle iron is usually available and inexpensive. However, this alternative requires the skills of a welder and a carpenter.

A more common and relatively inexpensive design is "nut and bolt" shelving. This involves punched steel angles for the uprights, and pressed sheet steel shelves. These units do not make use of local materials in a developing country, but their advantage is that the purchased steel angles allow shelves to be repositioned and extra shelves to be added at any time. The components are fastened together with nuts and bolts and a unit made of four uprights and five shelves can be assembled ready for use in 20 minutes. The angles themselves have a variety of other uses. They can be used to construct racks, benches or struts to strengthen the shelf units themselves.

Cost comparisons must take local circumstances such as shipping costs and customs duties into consideration. However, a simple check in England revealed that, for the standard shelf unit described above, nut and bolt steel shelving is cheapest. A similar iron angle unit with chipboard shelves cost 25 percent more for materials alone without the welder's time. A compartmentalized eight-bin unit of the same size cost 63 percent more. Nut and bolt shelving has further advantages in that standard components such as drawers and lockable cupboard doors can be added for additional security, and shelf dividers, lips, backs and sides are all available as bolt-on extras.

Standard catalogues offer a variety of shelf sizes. Shelf length is commonly 1 m with depths varying from 300 mm to 1 m. The usual depth is 300 mm for single-sided units and 450 mm for double-sided, free-standing units. Shelves of these sizes can carry loads of 115-150 kg. Heavy-duty versions can handle up to 170 kg without reinforcement if the load is evenly distributed. Some manufacturers offer steel channel reinforcement which can be added inside the rolled lip of the shelf, increasing capacity to 550 kg.

Parts may be separated on shelves by steel dividers; some shelves are sold ready punched with holes for fastening dividers. In most cases, however, this is an unnecessary expense since boxed parts really need no separators and small or unboxed parts can be kept in containers on the shelf.

The most convenient container for small parts is a cardboard box available from commercial storage equipment companies, or from any reputable local manufacturer of packaging materials or paper goods. The most suitable size is 250 mm long, 105 mm wide and 105 mm high. The boxes arrive folded flat and are assembled as needed by bending up the sides and ends and locking them in place with fold-over flaps. No glue or staples should be needed. The part number can be written clearly on the end of the box for easy identification.

A local paper goods manufacturer may charge for making the die needed for cutting and creasing the cardboard blanks, but even then the price should not be more than a few cents per box. They are thus cheap enough to be disposable. This type of box is also available from equipment suppliers and although they cost two to three times more than those produced locally, they are usually made of oil and grease-resistent cardboard, and come in a variety of sizes.

A more durable alternative is the plastic parts bin, available in a variety of sizes and colours. Some can be fitted with interior partitions; some can be stacked and most of them are open-fronted with a lip to keep the parts from rolling out. They can be placed on a shelf or be clipped to a louvred panel fastened to the wall. This is a good option for utilizing space which is too small for a regular shelf unit. The price of a plastic bin is about ten times that of a cardboard box of similar size.

There is a wide variety of metal or plastic boxes, bins, trays and containers for larger parts or bulk storage of larger quantities of small parts such as nuts and bolts. There are also wire baskets which can be stacked, steel mesh containers suitable for forklift handling and special racks for tyres, drums or oddly shaped goods.

Carrying parts in and out of the store is easier if they are grouped in batches as ordered and placed in a metal or plastic tote bin. Again, a wide variety of these is available commercially. In addition to equipment manufacturers, mail order companies publish annual, priced catalogues featuring a wide range of items.

5.4 Mechanical handling equipment

Mechanical assistance, necessary for moving heavy parts, varies from a simple trolley to an expensive forklift truck. A popular method for moving small quantities of parts is a tote bin on legs, or a pallet with a hand pallet truck. This is a device with two load forks which slide under the pallet. A manually-operated hydraulic cylinder then lifts the forks and the bin, which can be manoeuvred as required. Different types of hand pallet truck are available which can elevate and stack a load or place it on a vehicle. Most of these trucks have very small wheels or rollers and are only suitable for use on a smooth concrete floor. If loads must be carried over rough surfaces outside the store, a large-wheeled trolley should be used. There are also electric pallet trucks for the store, for frequent handling of heavy loads which vary from small units to full-scale forklifts.

For the very large store, special types of forklift can be bought for use in narrow aisles and to load pallets sideways into storage units. There are also special picking machines which carry the operator on a loading platform with all the machine's controls, so that stock can be picked from anywhere, at any height, along a row of storage units.

In a small store there is no need for such sophistication. However, heavy loads must be handled when large consignments of parts arrive and when heavy single parts, such as engines, must be moved. Where there is no forklift, a 1.5 tonne capacity chain hoist is relatively inexpensive to install. It can be hung from a trolley running on a rail bolted to the ceiling and can be used to carry heavy loads in and out of the store. The store ceiling may need reinforcement if it is to support a chain hoist rail. Some of the floor space under the hoist rail should be designated for storage of large unit loads.

Apart from regular maintenance tools, including an industrial vacuum cleaner, little other equipment is necessary. There should be tables and benches on which parts can be laid out for checking after arrival or before despatch. Where parts are sent long distances a packing table with a holder for rolls of wrapping paper and a case strapping tool will be useful. These are inexpensive items and can be ordered through mail order catalogues.

5.5 Office equipment

Office equipment for the parts department should include the normal range of office furniture and machines such as typewriters, calculators, cash tills and a photocopier if these services are not provided by a central unit in the organization.

A microfiche reader will probably be necessary for the counter. Some of these are specific to one type of fiche. Where a variety of parts from different manufacturers is handled, a study should be made of the different types of fiche they produce. It may be possible to avoid buying several readers if one can be found with a selection of lenses. Small, battery-powered microfiche readers can be valuable for occasional use in the manager's office or elsewhere in the store but are not suitable for counter use.

If the workshop service counter is any distance from the retail counter, a microfiche reader may be needed there as well.

5.6 Stock card index

Although stock record cards are rapidly being replaced by advanced computerized inventory control systems, in many parts departments they are still regarded as a system of the future.

The stock card index is the most important office item in the parts department. The most efficient systems involve cards in a cabinet, arranged so that part of each card is exposed. On this visible strip is written the part number and description. Normally the visible edge of each card - the portion that is handled as the cards are flipped over for inspection - is held in a transparent plastic protective sleeve.

One of the most popular systems uses 21 cm x 15 cm cards, held in drawers in sets of about 60 cards to a drawer. If there are 15 drawers in a cabinet, the cabinet holds about 1000 cards. The cards may be folded over a wire retainer or slotted into holders which are locked into the drawer. The important point is that the cards are removable for replacement purposes but are not loose in the drawer.

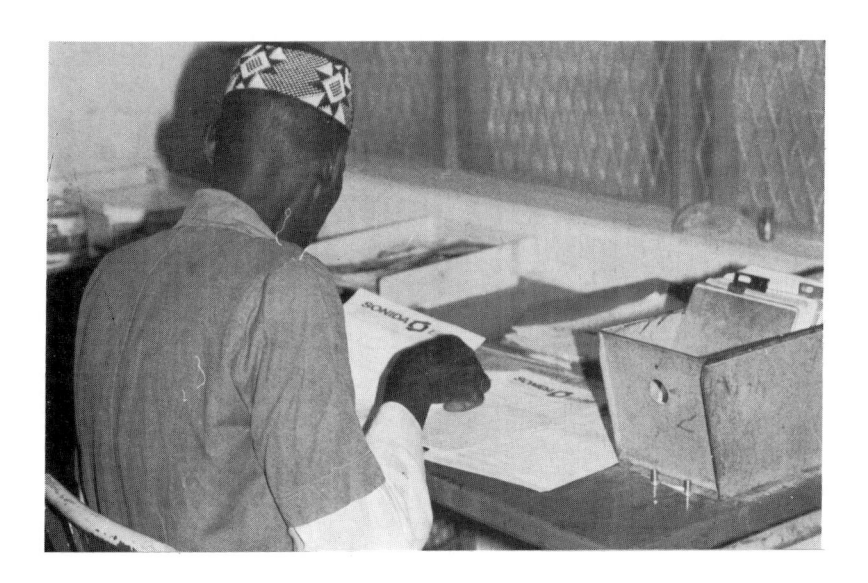

Another popular system uses larger cards stacked vertically in a series of trays. Notches on the cards secure them so that they overlap, leaving only the strip bearing the part number and description visible. To enter information on these cards, however, they must be removed from the trays, whereas in the system described in the preceding paragraph, individual cards are not removed and are thus less likely to be mislaid or misplaced.

Different types of card are available from stationery companies or from the designers of the various systems. Cards may also be printed locally at very little cost, with the design and language characteristics necessary for an individual operation.

Two main information systems are used. In the first, the parts movements and a record of orders are shown on the face of the card in simple chronological sequence. Both sides of the card are similar and only one type of card is needed. This system places all information on one card. When it is filled up, some information must be transferred to the new card, ensuring that every item on the card is up to date. This system is more suited to the small store.

In the second system, movement information is recorded on one side of the card or on one of two adjacent cards. Order records are kept on the second card or on the back of the first card. The two-card system is efficient because order information takes up less space and a longer history can be kept on one card. The movement record card is used up more quickly and carries more transient information. On the other hand, both cards must be viewed for a complete picture of the situation. It is probably the best system for a store where movement is rapid and movement cards are used up quickly. It also allows more space for recording sales history analysis on the order card.

The arrangement of cards in the drawers is important, since the objective is rapid information access. The cards should be in numerical sequence for a given set of parts, but the composition of the set must be decided in advance. Generally, all parts from one manufacturer will form a set, but the parts numbering system does not always divide parts into the most useful sub-sets. It is often necessary therefore, to divide the cards into groups representing, for example, all engine parts, all transmission parts, all axle parts, etc. They can then be arranged in sequence within these sub-sets and related parts are more easily located on cards near one another.

When determining the order of the cards - or when considering any aspect of a machine - it is useful to work from front to back: from the front axle to the engine, the clutch, etc. A systematic pattern will emerge from this process, making it easier for staff members who work on the cards to find them rapidly.

5.7 Card design

Card design is dictated by the type of information required by the operation, but there are some features common to all card types. There should be seven columns for movement information, six if the order record is on a separate card. Normally, on a 21 cm x 15 cm card there is room for two sets of columns on each side of the card.

The columns should be headed:

Date: month and day; the year can be shown at the top of the column or at whatever point a year's records start.

Reference: the invoice or requisition number on which parts are issued or the order number for both orders and arrivals.

Order: the quantity ordered on the reference number and date in the previous two columns.

Received: the quantity received on the order number in the reference column. If parts are returned by a customer, the reference will give the credit note number.

Issued: the quantity issued against the invoice or requisition number in the reference column.

Inventory Record Card

Supplier:								Address:											
Date	Ref.	Ord	In	Out	Bal	Cost				Date	Ref.	Ord	In	Out	Bal	Cost			

Replaces	Replaced By	Before S. No.	After S. No.	Cat	Min
Part Number	Name	Vehicle	Assembly	Location	

Balance: the quantity remaining in stock. This does not include parts on order. It is important that the stock balance shown on the card is always the same as the number of parts in the bin.

Cost: the landed in-store unit cost should be shown in this column. This is important even for non-commercial organizations since it is the basis for control.

Some card designs include a set of boxes to show each quarter's sales for four or five years. This is important data but the 15 cm depth of a movement record card only allows enough space for about 70 records of movement, and it is difficult to justify using this space for historical records which, to be significant, would have to contain many more than 70 movements. For a small system using only the movement record card, it is easier to total the issues at the end of each quarter and note the figure in the issues column.

Other important information that should be shown on the card includes:

Part number: prominently displayed at the bottom right-hand corner.

Description: equally prominent and corresponding to the standard description in the manufacturer's catalogue.

Location: where the parts are found in the store.

Machine: the tractor, vehicle or implement on which the part is used (possibly more than one).

Assembly: where on the machine the part is used (engine, clutch, etc.).

Price: the retail price of the part. Changes can be made by writing the new price on a self-adhesive label and placing it over the old price.

Above the bottom line, another series of boxes should show supersession information. This is essential for the proper recording of part number changes when a new part replaces an existing one. One box is needed for the old part number (on the card used for the current version). Another is needed for the new number (when a part has been replaced). Thus a card may show three part numbers. One, belonging to that card, represents a part in an intermediate position in a series of modifications and may register some remaining stock. It will refer to the part which the one on the card replaces. Whether or not there is stock left of the old part the part number is useful since it may appear on customer orders utilizing out-of-date information.

The card may also show a new number for the next stage in the modification series. No further orders should be placed for the old part. All future activity will take place on a new card representing a new number (but cross referencing the old one). These two boxes are labelled "Replaces" and "Replaced by". Two more boxes on the same line, labelled "Before Serial No.:" and "After Serial No.:" indicate machine models on which the part can be fitted.

The card should also have a box showing the order level or "minimum" stock at which a new order should be placed. If the simple order guide chart is used, the part can be categorized according to its movement rate and value, so it is useful to have a box on the card show this information.

The card should also include the name and address of the part manufacturer or supplier. If all the parts in a certain category come from the same supplier, it is not necessary to repeat the information on all the cards. However, there will be cases when parts such as electrical or fuel injection components can be purchased more easily from specialist suppliers and it helps to have a record of this information on the card.

A supply of coloured markers is necessary to flag cards on which some action is required. For example, a red marker may indicate the need for an emergency

Order Card

Supplier 1:					Address:						
Supplier 2:					Address:						
Supplier 3:					Address:						

Date	Reference	Sup	Ord	Recd. Qty	Recd. Dte	Cost	Recd. Qty	Recd. Dte	Cost	Recd. Qty	Recd. Dte	Cost	Bal	Min

Replaces	Replaced By	Before S. No.	After S. No.		
Part Number	Name	Vehicle	Assembly	Location	

Movement Card

Date	Reference	In	Out	Balance	In	Out	Balance	Replaced By	Before S. No.	After S. No.	Name	Vehicle	Assembly	Location
								Replaces			Part Number			

stock order; a green flag may signal that the stock is down to the normal order level. There are various types or markers available from commercial suppliers. Some are metal clips which attach to the plastic card protector sleeve. Others are plastic strips which can be inserted into the sleeve.

In a store carrying many fast-moving parts, an alternative is to use a separate order record. In some card designs, this information occupies the back of the movement record card. Since this may be replaced fairly frequently, it means that historical sales records must be copied onto a new card each time one is replaced.

It is better, therefore, to use a separate card, filed in an adjacent slot in the drawer, for sales analysis records and calculations of order level and economical order quantity. This card will need columns for details of orders and the dates on which the goods are received. The quantity received is entered on the movement record card and when the order is complete, it may simply be crossed off the order record card.

VI. STOCK PLANNING

6.1 Service level and investment

Planning parts stocks depends on the machinery population to be serviced, its past history of parts consumption and the service level to be achieved. In an on-going situation, consumption records will help with planning, but since stock is for future consumption, past history is only an indicator.

Unless records have been kept of all parts requested in relation to those invoiced out, it will be difficult to establish just what the service level has been. A service level calculated by looking at back order situations is not good enough. If service levels are not recorded, the first indication of inadequate service may be discovery that the enterprise has acquired a poor reputation for parts supply. By that time much damage may have been done and the cost of recovery will be high.

The first step in planning is to state the objectives; the market and service level should be clearly identified. In a competitive market where prices have to be kept down, there is no point in planning an ambitious service level because the investment cost will be too high. On the other hand, a project that is sensitive to machinery availability may need a high service level, and the investment may be worthwhile. It must be decided where the balance should lie, and what policy will be adopted to deal with the demand for parts that will fall outside the stocking range.

A high turnover rate is said to reflect a low service level. Good stock planning, however, will ensure that a reasonable turnover of between 1.5 and 2.0 can still be achieved together with a service level of at least 85 percent. Above this level, the cost of investment and the rate of turnover become less economic with every percentage point achieved.

Low turnover does not automatically mean a high level of service. The low turnover may be caused by stocking the wrong selection of parts, in which case the investment cost will be high in relation to the satisfaction achieved. In industrialized countries, where the supply source is close to the outlet, lead times will be short and even out-of-stock situations may be remedied the same day. Under these conditions, a rate of turnover of 5.0 or 6.0 may not be unreasonable. A high level of service between 90-97 percent can be achieved in such circumstances with a much lower stock investment.

Once the service level and the intensity of investment have been decided, it will be possible to consider the type of parts to be stocked and the investment budget.

6.2 Stock budget

In previous chapters, it has been suggested that where the maintenance and repair costs of a machine are measured in relation to its depreciation cost, the original value of the machine will give a good approximation of the value of parts used over the machine's lifetime. In a mixed population this would produce a fairly even flow of parts, year by year. In a homogeneous population on the other hand, parts consumption would be limited to maintenance items in the first year or two. Consumption of wearing parts would gradually increase until, as the machine entered its final years, the need for major overhauls created a much heavier parts demand. At the end of a machine's life, when its replacement is planned, further overhauls may be regarded as uneconomic and only maintenance parts may be consumed.

A parts manager starting a project, or taking over a store, should spend some time on these calculations. The figures should be reviewed annually according to changes in the population. Parts for one model may be seriously understocked, while parts for others are overstocked. It is a difficult accounting exercise to identify stocks on a -model basis, although a computer could be programmed for this analysis.

Stock sections are usually segregated according to manufacturer or general machine type. Tractor parts may be separated, for accounting purposes, from combine parts or truck parts. Engine parts may be grouped together even if several manufacturers are involved. This analysis helps judge results and plan reorganization of sections performing poorly. Whatever the system adopted, it is still possible to analyze the machinery population by number, value, age, and probable parts demand.

If a machine, in a lifetime of eight years, costs to maintain and repair about the same amount as its purchase price then these costs may be spread as follows:

Year	Repair cost	Parts (60%)
1	5%	3%
2	5%	3%
3	10%	6%
4	15%	9%
5	15%	9%
6	20%	12%
7	20%	12%
8	10%	6%
Total	100	Average 7.5

A mixed population of machines may be serviced with an annual supply of parts amounting to 7.5 percent of their new value. If the population is not evenly mixed by number and by age, the stock will have to be adjusted to take account of any disproportion. The value of a machine, for the purposes of this calculation, is its new replacement value each year because each year new parts are bought at prevailing prices and it can be assumed that inflation has the same effect on parts and machines alike.

If in Year 1 the new machine costs $10 000, the predicted value of parts purchases in Year 5 will be $900 (9%). But in Year 5 if inflation is at 5 percent, the cost of those parts will have increased to $1 150, which is 9 percent of a machine costing, new in Year 5, $12 765. Therefore, the parts manager, calculating what should be in stock for a five-year old machine, should use the current price of the same machine (or one similar to it), for the calculation.

For example, let us assume that a truck fleet has five vehicles bought ten years ago, five more bought five years ago, and five new trucks. If the price of that type of truck today is US$20 000 and if it is assumed that it needs 120 percent of its value in repairs over a life of 12 years, then this year's parts requirement can be found from the following table.

Year	Repair Cost %	Parts (60%)	Cost/ Truck $	Number	Total $
1	5	3	600	5	3 000
2	5	3			
3	5	3			
4	10	6			
5	10	6	1 200	5	6 000
6	10	6			
7	15	9			
8	15	9			
9	15	9			
10	20	12	2 400	5	12 000
11	5	3			
12	5	3			
Total	120			Total	$21 000

In a second example, assume a project in its fifth year is using a number of tractors and implements for cultivating and harvesting a crop. The parts consumption rates and working lives of the various machines will vary, but the calculation is done in the same way.

Three tractors: current value US$12 000 each, and expected to need in Year 5, 15 percent of new cost in repairs (9% in parts). Each tractor therefore needs US$1 080 in parts: US$3 240 total.

Three disc harrows: current value US$3 000 each, and expected to use 400 percent of their new cost in repairs over a life of 8 years. This would be mostly discs and bearings renewed annually at a cost of

50 percent of the machine's new value per year (30% parts). Each disc harrow therefore needs US$900 in parts: US$2 700 total.

One seed drill: current value US$8 000 and expected to use 200 percent of purchase cost in repairs over 8 years, more or less evenly spaced in Years 3-8 (15% per year or 9% in parts). The seed drill will therefore need US$720 in parts.

One fertilizer spreader: current value US$2 300 and expected to use 200 percent of new cost in repairs over a life of 6 years, more or less evenly spread at 30 percent per year. (18% in parts). The fertilizer spreader will therefore need US$414 in parts.

One combine: current value US$50 000 and expected to use 150 percent of purchase cost over a life of 8 years, with 25 percent in Year 5 (15% in parts). The combine will therefore need US$7 500 in parts.

Total parts needed for these machines will therefore be US$14 574.

In the following year these values will change. The replacement price of the machines will be higher (taking into account inflation in parts prices), but in their sixth year some will need relatively more parts as they approach a complete overhaul.

By making such analyses, the parts manager can estimate the overall demand for parts and plan stocking policy on a logical base.

6.3 Types of parts: maintenance parts

There are five broad categories: maintenance parts; consumables; hardware; wearing parts; and non-wearing parts.

Thorough knowledge of the maintenance requirements of the machinery population served by the department will enable the parts manager to make a fairly accurate calculation of the stock requirements for maintenance. There is no reason why a 100 percent service level should not be maintained for parts in this category. This level should be kept up because machine owners will at least expect to be able to follow prescribed maintenance routines.

Maintenance parts include all items which are essential to routine maintenance of a machine. Filter elements are the first in this category. Some, notably oil filters, must be replaced at periodic intervals laid down by the manufacturer or the machine will suffer exaggerated wear. Some filters, especially air filters, can often be replaced or cleaned. Under good conditions cleaning, by washing in a solvent or by compressed air, can extend the life of the element but not indefinitely. Under arduous conditions where there is any appreciable quantity of dust in the atmosphere, replacement is the safer option.

Fuel filters are extremely important and manufacturers' recommendations should be followed scrupulously. If there is any doubt about the quality or cleanliness of local fuel supplies, the fuel filter elements should be changed frequently. The parts department should liaise with the service department to establish a local recommendation for fuel filter changes. Fuel cleanliness is affected by the storage and handling processes the fuel passes through on its way to the engine. If it is stored in a bulk supply tank close to machinery operations and is pumped directly into machinery fuel tanks, the possibility of contamination will be reduced but not eliminated entirely.

In some organizations, parts storage is just one aspect of the responsibilities of an overall "stores" function, which will include fuel and other supplies. Correct procedures for handling fuel can be thought of in parallel with parts procedures. Cleanliness and order is particularly important in storing liquids such as fuels, lubricants, solvents or paints.

Even where fuel handling is not the responsibility of the parts department, the parts manager should have a very clear knowledge of local fuel handling methods and habits and should take these factors into consideration when planning fuel filter stocks. Good fuel handling is also part of the advice the sales staff should pass on to the customers. Some problems may be anticipated if parts and service departments stock and market fuel filters with greater capacity, finer filtration or additional features such as water separation.

Other items needed for maintenance are joints and gaskets, which are likely to be destroyed or damaged during maintenance operations, and various types of protective chemical. For petrol engines, parts such as spark plugs and contact breaker points, are necessary for routine maintenance.

The protective chemicals referred to may be consumable lubricants, such as grease, but other materials are now available to enhance the reliability and durability of machinery. For example, special coatings protect battery terminals from corrosion; others protect electrical wiring from moisture. Special compounds clean cooling systems and others stop them from leaking. In some cases, cooling water is chemically treated by a dispensing canister which looks like a filter element and has to be renewed periodically.

6.4 Consumables

This category describes parts or materials that are consumed as long as the machine is working. Lack of availability stops the machine. The most obvious example is fuel. In many projects and organizations where fuel is not available from the neighborhood pump, it must be obtained in bulk and stored ready for use. It then becomes a stores item in just the same way as a piston or a wheel bearing.

The same is true for lubricants. The diverse types of oil and grease required for different machines - or even for different parts of the same machine - may be available from local suppliers, but in many cases they are not and have to be stocked. This may represent a considerable investment. Although some machines can use "universal" oils for several functions, engine, gear box, hydraulic system, etc., other machines demand special oils for functions where high pressure hydraulics, or heavily loaded bearings or gears are concerned.

While a machine will not run without fuel, it can usually continue for a while without changing the lubricating oil. The inevitable result of failure to keep up with necessary maintenance schedules will be increased wear. Thus, an engine will start to consume more oil and there will inevitably be an oil supply requirement. Planning for lubricant supplies is therefore just as essential as planning fuel supplies.

The diversity of lubricants required may be offset by extended maintenance periods. For example, in one popular make of car, the manual gearbox and final drives are "filled for life". Fill and drain plugs are provided but it is not anticipated, under normal driving conditions, that the oil will ever have to be changed. Under heavy driving in the tropics, it may be necessary to reconsider the standard instructions to take account of the effect of high temperatures on the stability of the original oil supplied in the machine.

Water is a consumable in the service context, in that it is required for topping up batteries or for diluting battery acid for the initial fill of a dry-charged battery. Where bottled distilled water is available, there is no problem but importing water is usually costly. An alternative is to install a distillation apparatus, or at least a water de-ionizer which works on the ion-exchange principle by passing the local water through a column of impregnated resin beads. The result is not pure water, but its electrical characteristics are improved to the extent that it is safe to use in batteries. Where local water is turbid, it should be mechanically filtered first. Together with the de-ionizer, a regular supply of replacement resin cartridges will be needed at a rate depending on the amount of water treated, and its original degree of ionic contamination. Most de-ionizers are fitted with battery operated conductivity meters so the state of the resin can be checked.

Where consumable chemicals for routine maintenance are available on the local market, the workshop will buy new supplies when they are needed. If they are not available locally, the parts department may be asked to keep a stock. Such materials may also be sold to other workshops, creating additional business.

These compounds may include chemicals that remove corrosion products from a cooling system. In hard water areas they may be essential to prevent overheating associated with scale accumulation in an engine. Sealing compounds may be stocked to plug small leaks which would otherwise mean replacement of an expensive radiator.

Other types of sealing compound can also be stocked by the parts department to facilitate workshop operations. Some are necessary for the correct location of bearings, or to prevent components shaking loose, and are recommended in the manufacturer's repair manuals. Others serve as an adjunct to a gasket when component surfaces are no longer flat or smooth. Compounds are available that can replace a gasket under some conditions, providing a leak-proof, non-hardening permanent joint between two components. While most are water-proof, they are not all resistant to oil or fuel. It may be worthwhile to stock small quantities of other types of adhesive for specific or more general applications.

Penetrating oils are needed for components which are stuck with rust or deposits of carbon or dried lubricants. A wide range is available in aerosol or liquid form. Sometimes such oils protect against future corrosion; some have additives which ensure future lubrication. A selection will prove useful to the workshop. Graphitic or molybdenum disulphide lubricants for assembly and running in will also prove useful. The workshop will also need abrasives. Items such as valve grinding paste and emery cloth may be stocked by the parts department, and supplies of grinding wheels and discs may also be needed for workshop machinery.

Paint is another consumable which the parts department may stock for the workshop or for other departments of the organization. A management decision will be required before orders are considered, since paint and the associated thinners and cleaning solvents require special, separate storage in a fireproof environment and additional investment.

   Consumable parts and materials are not all used in the workshop. The ground-engaging parts of farm implements may also be considered consumables. Plough shares and harrow teeth are used up at the rate of so many per hundred hectares of land cultivated. Sometimes they can be re-built by welding but not all users have this facility. Without these new shares, ploughing becomes more power consuming and eventually impossible. There is, therefore, a good case for a high level of service in these consumable parts. Other parts in this category are discs for disc harrows and disc ploughs; ripper teeth; cutting edges for bulldozers and motor graders; and knife sections for combines and grass cutting machines.

   Tyres are usually available from specialist suppliers. If not, it may be worthwhile to stock tyres, but they do occupy a great deal of space. Even tyres for large agricultural and earthmoving equipment can be repaired and re-treaded, so it may be feasible to stock repair materials where stocking tyres is ruled out.

6.5 Hardware

"Hardware" describes a wide range of fasteners and associated items which are often bought in bulk. Machinery manufacturers are usually specific about their choice of fasteners, bolts, screws and nuts, for each application on a machine. This is because the technology of holding components together under stress is complex, and the design of appropriate fasteners has reached a high level of sophistication. The parts catalogue for a machine usually shows a part number against each fastener instead of simply describing it as "a 1/2" bolt", or "a 13 mm nut". Apart from technical aspects of fastener choice, this practice also brings the identification system for fasteners into line with the rest of the components in a machine.

Technical considerations are important but a great deal of standardization has been achieved. In machine production, there has to be a degree of rationalization, and if the same bolt is technically suitable for ten different applications, then it will be used for all ten. Design of fasteners relates to several criteria such as size, closeness of fit, thread form, tensile strength, hardness and national standards.

The size of a screw is not always critical in that a longer screw can often be used, (cut off if necessary). A smaller diameter bolt than the original may be used in an emergency but since it is also weaker than the original, it should be replaced at the earliest opportunity. Closeness of fit is also important, a smaller bolt will not work as the components may move in relation to one another and so provoke rapid wear, or upset machine adjustments. Closeness of fit is particularly important for parallel and taper pins and keys.

Thread forms have become standardized and should be selected to conform with the standards used on the machines to be serviced. Nuts must be selected to conform to the bolts in use. The British BSF and Whitworth thread forms have largely been replaced by either American (UNF or UNC) forms or metric types. These two standards are now the most common in machinery throughout the world.

Tensile strength in a fastener can be especially important where highly stressed components demand high tensile fasteners. Lower grades may disrupt critical assemblies and destroy expensive components. A differential assembly is a good example where, if the correct bolts are not used to hold the crown wheel on the differential casing, the casing itself and all the pinions, shafts, bearings, seals and washers within it, may be ruined. The tensile strength grade of a bolt may be marked on it, and a replacement should always be of the same grade. If there is no marking and the application is in a critical area, the manufacturer's original part should be used rather than a common bolt.

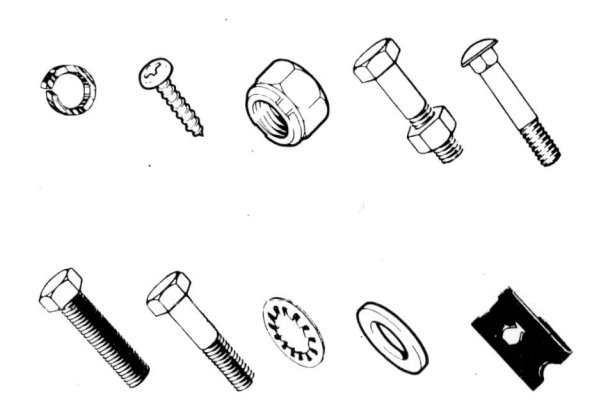

Tensile strength and hardness are critical in a shear bolt used as a safety feature on a machine. The wrong bolt may mean failure to shear under the appropriate stress and consequently damage to the whole machine. If a plough beam is expected to trip when the share hits an obstruction, the tractor driver will proceed at a pace determined more by operational economics than by fear of obstacles. At this pace, if the wrong shear bolt has been used and the plough fails to trip on striking an obstacle, the plough will be severely damaged and the tractor and operator may also suffer.

Some manufacturers' catalogues identify the type of fastener against each part number. Some manufacturers use a special code, through which the part number itself will identify the fastener. Where a manufacturer does not explicitly identify the type of fastener used against each part number, then sample orders can permit physical identification.

The aim of this identification is to allow the purchase of fasteners in bulk. Many items will be common, not only to several applications in one machine, but also right across the range of machines built by that manufacturer and possibly several others whose machines have to be serviced. The greater proportion are not specialized, and items such as common bolts, washers, split pins and rivets are better bought by weight than individually.

Fasteners, such as split pins and circlips, can be bought in boxes ranging from several hundred to thousands, in which there is a selection of useful sizes. These selections make a good starter pack. Some sizes will be more popular than others and these can be bought in bulk to replenish the stock.

Sealing rings and "O"-rings are not, strictly speaking, hardware but they may be bought in bulk or in compartmentalized packs of popular sizes as described above. Copper, fibre, or bonded sealing rings are a common requirement. "O"-rings are more critical since some have special design characteristics such as hardness or composition to resist special oils. Nevertheless, a boxed selection of popular sizes can solve many problems. "O"-rings can also be made from spliced lengths of rubber cord, and kits are available including cord and adhesive.

Specialist manufacturers of oil seals often produce cross reference catalogues which facilitate bulk ordering of popular sizes, but it is more difficult to identify the seal that fulfils the precise requirement . Dimensional tolerances between the seal, its housing and the shaft rotating within it, are critical. There may be a special requirement for multiple lips, for shielding on one or both sides, and for varying degrees of spring loading, rubber hardness or resistance to special oils. The greater variety in oil seal types, and the consequently reduced demand for each, makes them less suitable for bulk buying.

If there is a high proportion of hydraulically controlled equipment to be serviced, as is often the case on a construction site, hydraulic hose and fittings may usefully be bought in bulk. This may fit in with a workshop section specializing in hydraulic repairs or in the assembly of hoses which can be a profitable line to be marketed within, and beyond, the organization's own machinery population. This would require a selection of hose sizes and pressure capacities as well as an adequate selection of end fittings.

Selection of fittings can be complex since both male and female types with a variety of screw fittings, swivelling or otherwise, and various angles of approach will be required. Previous demand is the best guide, otherwise a careful study of several parts catalogues will have to be made.

The list of items to be included under the general category of hardware is long. The following provide a useful selection:

Set screws (threaded full length): UNF, UNC; diameters 1/4-5/8"; length 1-2 " (Note: a set screw that is too long can always be cut short.

Bolts (plain shank with end portion threaded): UNF, UNC, Metric; diameters 5/16-3/4", and 6-12 mm; length 2.5-4" plus 5" and 6" in 3/4"diameter, 25-70 mm. (Note: a bolt that is too long may be shortened if additional thread can be cut on the plain portion of the shank.

Plough bolts: according to type of implement serviced; most plough bolts may be bought in bulk.

Shear bolts: according to type of implement serviced.

Carriage bolts (round head, square neck): where timber framing or cladding is much used.

Nuts: to fit above bolts - normal hexagon nuts, lock nuts and self locking nuts.

Washers: flat, spring, internal and external shakeproof, in diameters from 1/4-3/4" (6-19 mm).

Sealing washers: copper (6-15 mm), fibre (6-27 mm), and bonded (12-20 mm).

"O"-Rings: 5/16-3/4" (7.6-19 mm) in 1/16" (1.78 mm) section; 7/16-1.75" (11-49 mm) in 3/32" (2.62 mm) section.

Cotter pins (split pins): 1/16" x 1/2" (1.58 mm x 12.7 mm) to 9/32 x 2.5" (7.14 x 63.5 mm).

Clevis pins: 5/16-1/2" (8-12 mm) up to 2.5" (63.5 mm) long.

Roll pins (tension pins): 1/8-3/8" (3-10 mm) and up to 3" (80 mm) long.

Circlips (internal/external): 7/16-1.5"(11-38 mm).

Hose clips: 3/8-1.5" (9.5-12 mm) to 2.3/8-3.1/8" (60-80 mm).

Exhaust pipe clamps: 1.75-2.25" (44-57 mm).

Hydraulic hose clamps: 1/2" (12 mm) and 3/8" (10 mm).

Wire rope clamps: 1/8-9/16" (3-14 mm).

Hydraulic linkage pins and linch pins: according to equipment to be serviced.

Grease nipples: 1/8",1/4" and 5/16" straight, 45 & 67 deg.; 3 mm, 8 mm and 10 mm straight, 45 & 90 deg.

Electrical connectors: assorted sizes, crimp-type lugs.

Self-tapping screws: assorted sizes.

Pop-rivets: 3-5 mm diameter; 6-12 mm shank length.

Wood screws and nails: where applicable.

Hydraulic hose and end fittings: according to the type of quipment serviced.

The range of hardware to be stocked will depend on the nature of the equipment to be serviced and the local availability of this type of item.

6.6 Wearing parts

Wearing parts require the greatest degree of skill in stock planning. Wearing parts may have a definable consumption pattern related to use of the equipment, but the rate of wear is usually relatively slow, and by the time it has reached a point where action needs to be taken, the machine may have changed hands and dropped into a different utilization pattern.

This applies particularly to engine and drive train parts. Where maintenance is good, a machine may work for its expected life span without an engine overhaul. This is one benefit of modern materials and lubrication engineering. In other cases, planned overhaul schedules make replacement of certain parts mandatory, but this is usually only in the case of expensive machines such as mining equipment where high purchase cost mandates a long life and repeated overhauls.

Here again, the manager should know how the machines are used and be aware of local practices concerning replacement and overhaul. For example, large-scale farmers may buy new machines every two years, in which case they would be unlikely to be faced with any overhauls, or to buy wearing parts. They may sell their two- year-old machines to small-scale farmers who expect to get another five year's use out of the machines.

They too may expect to re-sell the machines before they need major overhauls, but they will have to buy some wearing parts. Within the period of this second ownership, conditions may change. The owners may join the ranks of the new machinery users, and the old equipment may be relegated to secondary, or stand-by functions where the wear rate is much lower and the need for major repairs will diminish. On the other hand, they may, for economic reasons, be forced to retain the old machines longer than planned, and a complete overhaul may be the only way reliability can be assured.

Alternatively, in a prosperous region, most machinery users may use nothing for more than two or three years. There will be very little market for second-hand equipment there and it will have to be sold outside the area - possibly even overseas. Dealers in such an area would not stock wearing parts.

In the area receiving the used machines - especially if they are imported into a different country - there could be a large demand for wearing parts. The general principles for planning stocks of this type depend on the pattern of machinery use but there are a number of other considerations.

First, the cost of overhauling a machine is always high; in addition to the mechanic's time, the real or theoretical costs of not having the machine in use must also be counted.

Whenever an assembly is repaired, there will be time-consuming ancillary operations, such as cleaning, dismantling surrounding parts and building them up again, and then testing. It is a good idea to do whatever is necessary to prevent a repeat visit to the workshop for any reason associated with the original repair job, or any other repair that might be anticipated by proper examination of the machine in its dismantled state.

For example, if an engine is dismantled to replace a broken piston ring, the likelihood of having to dismantle it again in the near future to replace other rings, pistons, liners, or all three, must be weighed against the cost of enlarging the original job. If, after the original minimal cost repair, the machine has to go into the workshop for a similar job, the cost of being out of use, transport to and from the workshop and the ancillary costs of dismantling, re-building and testing will all have to be repeated.

The cost and time lost waiting time for parts, and possibly for a vacant spot in the workshop's work schedule, may be more than the repair. So once a machine is in the shop, a general assessment should be made. Providing the cost in parts is not excessive in relation to the value of the machine, and if the additional time is not out of proportion, the machinery user will usually prefer the replacement of all components nearing the end of their reliable life - or at least all components within the major assembly originally giving rise to the breakdown.

Stocks should be thought of in sets. Piston rings should be stocked in sets, together with pistons, liners and crankshaft bearings, and all the necessary gaskets and seals to enable a complete job to be done. An occasional breakdown on a new engine may mean that only one part is replaced. Ordering procedures should thus be geared to rebuilding sets.
The engine set may be thought of as several sub-sets. A cylinder head, for example, with valves, guides, springs, inserts, seals and gaskets, is one complete and distinct area for service problems and specific repair jobs. A crankshaft complete with its bearings, thrust washers and seals, is another useful sub-set, often sold in kit form by the manufacturer. However, if a crankshaft or its bearings are replaced, the pistons and cylinder liners should also be checked.

A crankshaft may be repaired two or three times if provided for in the manufacturer's design, and if local facilities are available to re-grind it accurately and with the requisite finish. Each time it will need new bearings of a specific under-size dimension. In stocking parts for the bottom end of the engine, design characteristics and local crankshaft re-grinding facilities should be taken into consideration. The quantity of main and connecting rod bearings should be divided into the required proportions of standard and undersizes suggested by the age and state of the engine population.

Another sub-set related to the engine includes parts for the fuel pumps and injectors. It makes little sense to overhaul a fuel pump without making sure that the injectors are in good condition. Parts such as steering ball joints, wheel bearings and seals, gearbox and final drive components, wear at varying rates which can only be judged from experience. For example, a truck fleet in mountainous country may suffer a great deal of gearbox wear, sometimes to the extent that a service exchange programme for gearboxes could be beneficial. However, gearbox wear in farm tractors is usually less, which would therefore reduce the number of overhauls needed during the lifetime of the tractors.

Stock planning for these miscellaneous wearing parts is a matter of observing trends and anticipating future demand. In a large machine population, this may be done fairly accurately as long as the dynamics of the population are kept in mind. A small population of less than fifteen units of the same model is more difficult to plan for. In this case, it is probably better to gamble on needing a few inexpensive parts and relegate the expensive ones to the "order on demand" category, at least until experience has been built up regarding wear rates.

6.7 Service exchange units

A fuel pump overhaul is an intricate and lengthy process requiring specialized machinery. To enable the engine to be returned to service rapidly, the complete pump is now usually replaced. Likewise, a complete set of injectors may be installed without waiting for the original ones to be repaired and checked. The machine then returns to work with the least possible delay, and the defective pump and worn injectors are retained by the workshop for overhaul when it is convenient and when all the parts have been received. These assemblies are then returned to store to be used next time a similar problem arises.

This process of sub-assembly replacement is most common with fuel systems and electrical components such as starters and alternators, but can also be extended to complete cylinder heads, clutches, hydraulic pumps, valve blocks and rams, complete gearboxes and even engines.

In the case of the engine, it may be either a complete unit with all accessories, or what is known as a short, or half engine in which the replacement assembly consists of only the cylinder block together with the main moving and wearing parts, including crankshaft, camshaft, pistons, liners, bearings and seals. Auxiliary sub-assemblies are then added to it from the original engine, either as they are or as replacement units themselves.

For machinery users, sub-assembly replacement means their machines return to work in the shortest time, and often for the least cost. Initially, users may object to paying for a complete assembly if they know only a small part of the old one needed replacement. However, the labour and transport costs of repairing the original assembly can be greater, and the machine may be out of service longer, especially if parts are not available locally. In the case of a large machine such as a bulldozer, replacement sub-assemblies can mean the difference between a job done rapidly and simply in the field, and the cost and time required to transport the machine to a workshop.

The use of complete sub-assemblies, or service exchange units, as they are commonly called, is now widespread and most machinery users accept the benefits as long as they can see that it makes for logical savings. An additional benefit is the guarantee that usually comes with the service exchange unit, especially if it is a factory re-built assembly.

In industrialized countries, it is often the original equipment manufacturers who undertake the job of reconditioning defective assemblies, either in their own factories or under closely supervised subcontracts. Thus, local repair workshops become replacers of complete sub-assemblies. They dismantle less, rebuild quickly, and actually repair little. They need stock few small parts, need less specialized workshop equipment and can handle more customers for a given size of premises and work force.

In less industrialized countries, it may not be practical to return defective assemblies to a manufacturer for rebuilding. Apart from the freight cost, the procedures for re-exporting worn parts of imported vehicles are often extremely cumbersome. Nevertheless, it is often possible to import factory reconditioned units, either by arranging with an exporter to find, in the country of origin, the requisite number of scrap units for return to the manufacturer, or by paying the premium demanded by the manufacturer for the non-return of the defective original unit.

A project, organization, or workshop can operate its own service exchange scheme. With proper planning and close consultation between the service and parts departments and the financial controller, a flow of reconditioned units can enable machines to be returned to work quickly and enable the workshop to even out its workload. Problems are resolved quickly, and the defective units can then be rebuilt under good conditions with adequate time, proper supervision, and when all needed parts are available.

From the point of view of parts planning, it must be decided precisely what units should be included in the scheme; in the early stages it may be better to restrict the number to one or two until the procedures are established.

It must also be decided who will stock the reconditioned units. If they are to be returned to the store, a value has to be put on them. While they are in the workshop they have to be clearly identified and not mixed up with other similar units removed from machines under repair. The job cards relating to their repair should show that they have to be returned to the store on completion. When they are returned to the store the accounting processes should keep track of each unit and its history. This may be done by giving each unit introduced to the scheme a special serial number, e.g. a service exchange starter motor may be numbered "SES/1". When that is exchanged for a defective unit, the unit brought in will be "SES/1/1". When that one is exchanged, after repair, the next one in the series will be "SES/1/2". The second new starter introduced to the scheme will be "SES/2" and will be resurrected as "SES/2/1-2-3-", etc.

The third consideration is pricing. After the scheme has been operating for some time, it will be possible to calculate the repair costs and the number of times the cycle is repeated for each new unit brought into stock. Theoretically this can go on indefinitely, but eventually a unit will be lost or stolen, or be damaged beyond repair. Based on the number of times a unit cycles and the repair costs, a standard price can be calculated. This will yield some profit to the workshop, a normal profit for the sale of the repair parts used and a share for each cycle of the original cost of the new unit that initiates the process.

For example, a starter motor may initially cost about US$250. It may be replaced because of a faulty solenoid worth US$65, or a worn drive assembly valued at about US$35, or simply for replacement of worn brushes at about US$3 a set. Labour costs will add to the price of each repair job. The total process over five cycles can be seen from the table below which shows that US$235 will have been charged for labour at full price. Since this repair work will have been done during slack periods, the labour cost can be regarded as the minimum standard cost and the US$235 will yield standard profit to the workshop, as shown on the next page.

```
Price of new starter....US$250.00 (net cost US$125)

Repair 1.  Replace solenoid............65.00
           Labour......................50.00
Repair 2.  Replace drive assembly....  35.00
           Labour......................50.00
Repair 3.  Replace solenoid & brushes..68.00
           Labour......................50.00
Repair 4.  Replace brushes............. 3.00
           Labour......................35.00
Repair 5.  Replace solenoid............65.00
           Labour......................50.00
Total cost.........................US$721.00

Standard price to customer......US$144.20 (US$721.00)
                                             ─────────
                                                 5
```

The parts cost of US$236 will also represent full retail price, and thus a standard profit margin. The unit exchanged for Repair No. 5 will be repaired and sold again. The overall profit is less than would be made by selling five or six new starters, but the machine users are better off because, while paying more than they would for minor components, they are getting completely overhauled and reconditioned units for much less than the price of new ones, and without waiting for a repair job to be done.

The value of the reconditioned unit returned to stock would be, for the purposes of accounting and financial control, the share borne by that unit of the cost of the original new unit, plus the price for the repairs.

Occasionally service exchange units are all charged out at new price, but this is unusual. Sometimes the users are simply charged for the repair of their own original assemblies, plus a share of the price of the new units. This may seem more equitable especially when a minor component is at fault, but it means that the price cannot be determined until the repair is actually done. If the unit is exchanged at the height of the busy season, the repair and the invoice may have to wait for weeks. This leads to lax accounting, poor credit control and an exaggerated investment in "work in progress".

The use of service exchange units should be encouraged because it facilitates stock planning by taking the urgency out of the parts demand. If sufficient new units are brought in to build up a stock of continuously cycling exchange units, then the parts to repair them can be brought in to correspond more closely to the actual repairs planned, and timed to coincide with the out-of-season exchange unit repair programme.

6.8 Non-wearing parts

A significant number of parts are required from time to time to repair damage due to accidents rather than wear.

In an engine using cylinder liners, for example, the cylinder block will not normally need replacement during the life of the machine. Blocks will be replaced only if they are broken in a crash or by frost or by a broken internal component such as a connecting rod or if mating surfaces are damaged by erosion due to a faulty cylinder head gasket.

In an engine where the pistons run in bores cut directly in the block, it is usually possible to regrind the bores at least twice to accommodate wear, but if scoring or ovalization goes deeper, the block will need replacement. Major non-wearing castings such as cylinder blocks, gearbox housings, or axle casings, should not be stocked. If accidents occur, the part can be brought in under emergency procedures.

A benefit of the service exchange programme is that an exchange engine or half-engine, is built up around a spare cylinder block, mainly because this saves the time required to build the new main moving parts into the old block. In an organization using service exchange engines, there will always be a replacement cylinder block available, and a new one, to return the exchange scheme to operation, may be brought in without the need for emergency procedures.

The only non-wearing parts worth stocking are those which can be easily damaged, such as radiators, or those liable to pilferage, such as spare wheels.

6.9 Suggested stock lists

Stock planning has been much influenced by manufacturers' guidance on the support their equipment needs. However, operating patterns vary so much from area to area that the manufacturers' suggested stock lists must be subject to close scrutiny.

Sometimes the manufacturer's service staff provide the suggested list, or at least help compile it. In this case there may be exaggerated enthusiasm for supporting the workshop with a high level of service without corresponding care for the economics of the investment suggested.

Manufacturers' suggested stock lists may be welcomed as a starting point but they should be reviewed in the light of personal experience and local practice. Anything that looks like a non-wearing item should be eliminated. Sets should be examined for completeness. Items such as hardware, which may be common to other machines already serviced, should be examined with a view to rationalizing as much as possible. Parts which may be more conveniently bought locally may be eliminated unless price differentials, or a reliable prediction of volume sales, suggest otherwise. Rather than investing US$40 in a bearing which may be needed once per year, and is available on the local market, it would be better to put the money into a set of injector nozzles which can be sold five or six times during the year.

6.10 Stock in relation to machine type

Up to this point, discussion on different categories of parts has related to the type of part rather than the type of machine. In the following paragraphs some consideration is given to the varying types of machine to be covered.

6.11 Motor vehicles

In this category both automobiles (cars) and trucks (lorries) can be included since the basic construction is similar and in most cases, they both run on roads. Four-wheel drive vehicles can also be included. In most countries, passenger vehicles are much more widespread than either trucks or machines. This has led to the growth of a range of support services, from the official distributor through a network of dealers and small workshops in most places. The range of parts sources is usually even wider. In addition to the official dealer network, there is often a section of the local market devoted to the sale of automotive parts. However, in many cases, particularly in rural areas, manufacturer's support is not adequate and there may be a case for stocking parts.

Maintenance parts are essential. Depending on the number of vehicles in use, stocking sets of engine overhaul parts, even a half-engine, may be worthwhile. Other parts normally required include wearing parts for steering, brakes, gearboxes, axle shafts, drive shaft joints, clutches, windscreen wipers, electrical system components including both light bulbs and complete lamp assemblies. Shock absorbers, springs and engine-mounting rubbers also suffer under the conditions in which these vehicles are used. It is not sufficient to simply include a pre-determined supply of parts with new vehicles. Parts support should be integrated with a permanent parts operation providing for re-stocking and an adjustment of stocks to anticipated requirements.

Truck parts follow a similar pattern to the four-wheel drive automobiles if road conditions are anything other than ideal. It is not unusual, in large countries with poor roads, for trucks to need a complete overhaul of brakes, springs and shock absorbers after each round trip between the coast and the interior. The delivery of new trucks by road from the coast to inland bases often causes sufficient wear and tear to need a complete check and some serious repairs.

For trucks, maintenance parts are essential and wearing parts for drive train, steering, brakes and suspension, will also be needed. Hilly terrain will produce a need for gearbox and final drive parts. Engines may suffer, depending on the quality of fuel and lubricants, the regularity of servicing and the amount of dust in the air.

In a commercial situation, a truck is only earning while it is moving; certainly for 10-12 hours per day, and probably for 24 hours per day. Minimizing down-time is extremely important in a commercial truck fleet, or for a project transport system that is designed to ensure timely delivery of inputs and outputs. For this reason, the use of service exchange units should be developed to the point where very little repair work needs to be done actually on the vehicle.

Rapid removal and replacement of complete assemblies will ensure that time in the workshop is minimal. When trucks break down away from home, a service vehicle equipped with tools and the appropriate service exchange assemblies, will be able to repair the vehicle quickly. Repairs to the defective assembly can then be done in the workshop with proper equipment, in good conditions, rather than at the roadside.

6.12 Tractors

Farm tractors have become more complex in recent years and now often incorporate electronic circuits in the hydraulic control system, cab air conditioners, multi-range, hydraulically operated or synchromesh gearboxes, and a wide range of options for external implement control.

For developed countries, such developments are welcome where their increasing complexity can be supported with parts and service technology. In developing countries where such support is not yet available, one option is to buy tractors from the less industrialized countries such as India or Brazil, where the old, simple, models are still made.

Parts procurement may be more difficult as lack of complexity in the tractor may be matched by lack of refinement in parts organization, coupled with language problems and less direct transport connections. However, these problems can be overcome with good communications and adjustment to lead times.

If it is decided to switch from a traditional to a new supply source for similar models, there may be some overlap between the parts from the two sources. This is seldom complete however, and the model from the new source should be treated as completely different in its parts composition until experience proves otherwise.

If there is no choice other than the most up-to-date and complex model, the manufacturer must train the user and provide local support services to look after the new complexities.

Often, the sophisticated control system components cannot be repaired without equally complex tools and diagnostic equipment which are rare outside specialized industrial repair centres. In this case, the problem can be resolved by replacing the whole unit. If it can be sent away for repair so much the better, but in many countries, a defective electronic circuit board, or electro-hydraulic control valve, will simply have to be scrapped. The cost may be high, but the improvements in operator productivity brought about by the advanced technology will probably make up for it.

Advanced technology in a tractor necessitates stocking a selection of the special assemblies involved. They may be selected on the basis of whether their absence prevents the tractor from working. For example, one defective control valve in a bank of valves may not matter if the implement can be operated from an alternative valve. The electronic function in a hydraulic system may not be vital to basic implement operation, only to a refinement of it. The air conditioner is not essential to the operation of a tractor although operators may need persuasion to work without it. On the other hand, a defective electronic ignition system in a petrol engine will prevent the engine from being used and there is no alternative to stocking and replacing these components.

Parts stock for the traditional tractor should include, in addition to essential maintenance items, steering joints, wheel bearings, brake linings, clutch components, hydraulic system components (especially linkage parts), electrical system components, batteries and tyres. Engine parts such as pistons, rings, liners, bearings, seals, valves and crankshafts will also be needed if overhauls are expected, and for emergencies such as lubrication failure. Water pumps (or pump repair kits) will be needed in areas where hard or polluted water causes scaling, corrosion or bearing and seal problems.

Stocking of service exchange units will depend on the intensity of tractor operation, and on how easy it is to bring the tractor into a workshop. Where tractors operate more than 1 000 hours per year, or where they do not return to a base workshop for long periods, the use of service exchange units should be encouraged.

A tractor, like a truck, is only earning while it is working. In spite of the seasonal nature of its work, it is just as vital to have the tractor operational when it is needed and parts stocking plans should be viewed accordingly.

6.13 Cultivation implements

The stocking requirements for cultivation implement parts varies considerably. A plough, for example, may only need consumable items such as shares, points, landsides, coulters or discs. The rate at which these are consumed can be determined by the soil conditions. Some wearing parts such as coulter and disc bearings, will be needed, and accidents may result in an occasional need for a non-wearing part. Reversible ploughs have wearing parts in the reversing mechanism and these should be stocked since the plough will be unreliable once they become worn. Ploughs with safety trip beams will need shear pins because the lack of the proper pin or bolt will lead to the use of improper ones, and either excessive tripping or serious damage due to failure to trip.

Secondary cultivation implements, such as disc harrows or tined cultivators and harrows, will need consumable ground-engaging parts and, in the case of disc implements, bearings. The cost of maintaining a disc harrow during its life may be four times its new cost. It is important to stock these discs and bearings or performance can be drastically reduced. This applies especially to heavy discs used for primary as well as secondary cultivation. It is not unusual for cropping system to decline with poor growth and uneconomically low yields simply because lack of repair to the cultivation implements made it impossible to prepare the land adequately for planting.

Apart from consumable and wearing parts, cultivation implements depend on a good supply of hardware. The constant bucking and vibration caused by their passage through the soil loosen nuts. If they are not tightened daily, the implement will be damaged or fall apart. A supply of bolts, nuts and washers is therefore essential.

6.14 Planting equipment

Seed drills and planters are used only for short periods, but if they do not operate properly, gappy crops and reduced yields may be difficult to remedy as they do not become apparent for some time after planting is completed. This type of equipment must be checked thoroughly before use, and early enough to ensure that any parts needed can be procured before the machine is needed. Timeliness is vital in planting and a delay of only a few hours can mean a reduced crop.

Parts required for planting equipment are mainly rubber or plastic items, such as tubes and bushes, which deteriorate with time and exposure to sunlight and agricultural chemicals. The ground-engaging parts seldom get enough wear to require replacement unless they are used in very abrasive or stoney soils. Drive train parts such as belts, chains, bearings and couplings may also be stocked.

6.15 Fertilizer and manure spreaders (including combine drills)

The corrosive nature of the product handled by these machines, whether inorganic or organic, means that they usually last only five or six years. Machines with plastic hoppers may have a slightly longer life but these are more subject to physical damage.
A fertilizer spreader needs to be kept clean to reduce corrosion to a minimum. The parts needed are mainly belts, chains and bearings for the rotating parts.

6.16 Sprayers

The pump (with its drive) and the nozzles, are the most important items in a sprayer. Some pumps suffer more than others, particularly if dirty water is used in the spray mix. Although it may be strained, the strainer will not remove fine abrasive particles which will damage the pump. Nozzles will be eroded, especially if the water is not clean, and a range of nozzles of different sizes for different applications should be stocked. Apart from these items, hoses, hose clips, hardware of various types and valve seals will be needed.

6.17 Forage equipment

Forage crop cutters and various mowing machines will need knife sections and rivets as consumables. Wearing parts such as bearings and drive train components will also be needed. Balers are highly stressed machines and need maintenance and adjustment, but not, normally, large quantities of parts. Packer arms, particularly the disposable wooden type, and pick-up tines will be needed. Needles should also be stocked since they are frequently damaged, and when the twine eye becomes worn, the needle has to be replaced, otherwise the knotting function becomes unreliable. Knotter knives and wire grips will need replacement as wearing parts. Ram guides and slides on some models eventually need replacement. Rolls of twine or wire are essential consumables and should be stocked as balers will not work without them.

6.18 Combine harvesters

A combine works for a relatively short period each year but its task is a punishing one. A medium-sized combine can be expected to handle more than 50 tonnes of grain per day, (some of it twice), and an equal weight of straw. The grain may be damp and sticky; it may be full of dust and dirt; it may be abrasive - especially in the case of paddy. These factors produce accelerated wear and tear in the machine, and a combine can have a large appetite for parts.

Maintenance parts for the engine and hydraulic system are essential. Consumables will be limited to knife sections. A variety of hardware will be needed, especially nuts, bolts, circlips and pins. Wearing parts on a combine are also very varied. Drive trains, apart from that which moves the machine, include belt drives, with or without speed varying attachments, and chain drives which will need replacement. There may also be hydraulic drives with a need for pump and motor parts or service exchange units, as well as hoses and couplings.

Each combine has a series of conveying systems using slatted chains or rollers for crop input and elevation, and augers for grain handling. Grain augers are particularly hard on bearings and drive components. Grain is also handled by vertical elevators using a chain fitted with paddles. These and the sprockets and shaft bearings are also subject to wear. Grain is usually cleaned in a combine by an oscillating shoe filled with screens. The drive to the shoe, and the hangers on which it rocks to and fro, need frequent replacement on some machines.

In addition to the moving parts, there are sealing strips, curtains and deflectors which wear and suffer damage. Even sheet metal items such as elevator bottoms and auger flights will wear on a combine, especially if it is used to harvest paddy. Often these highly abraded areas have replaceable sections.

Wearing part stocks for combines should include all small bearings, bolts, anything that looks like a wearing or sealing strip, and a good selection of hydraulic hoses and couplings. Due consideration must also be given to the steering, braking and transmission systems of the combine. These machines may not travel far in the course of the year, but they are heavy and often highly stressed.

Modern combines have air conditioned cabs and electronic monitoring systems. Usually the combine can work without these devices but breakdowns are a nuisance. The air conditioner is more important for a combine than for a tractor, but its value lies in giving the operator a clean environment. If it is difficult to repair the cooling system, the fan unit will keep out most of the dust and provide some ventilation. Complex devices which cannot be repaired locally should be stocked as complete assemblies.

The main threshing components, the cylinder and concave, will normally need replacement at infrequent intervals unless there is an accident. Damage from rocks or metal objects taken in with the crop can often be remedied by straightening bent bars. However, where peg-tooth cylinders are used for rice harvesting, the pegs will need replacement and should be stocked.

6.19 Pumps and irrigation equipment

Parts for irrigation systems depend on the type of system in use. Most overhead systems use sprinklers which require frequent replacement since they become eroded by unclean water. Moving parts stick as they become encrusted with lime scale. Pipe lines may need coupling latches; they will certainly need sealing washers.

Whatever the type of system, its heart is the pump and its driver. Pump life depends largely on water quality. If the water contains abrasive particles, the impeller, bearings, seals and shaft will be liable to wear. Care is needed in stocking impellers; some are standard but many are machined to size to provide specific output for specific site conditions. When ordering impellers, therefore, the dimensions of the original and the pump serial number should be quoted.

The pump driver may be an engine or an electric motor. As long as connections and switchgear are in good condition, and the motor is protected from overload and the entry of water, there is usually no need to stock parts for electric motors. Where there are many pumps of the same type, it is preferable to stock one or two complete motors, and to have any that become defective rewound and overhauled by a specialist in this field.

Engines for pump drive applications are usually similar to tractor or truck engines and require the same type of maintenance and wearing parts. However, where irrigation is continuous, the engine may work day and night for weeks at a time, running at the same speed. While this type of operation is beneficial to an engine, it does require good maintenance. Particular care should therefore be given to maintenance parts.

The availability of pumped water may be more vital than the availability of tractor power or transport. For this reason, and because some engines often have to run unattended for long periods, they are more likely to be serviced under a preventive maintenance programme. Without waiting for a breakdown or a marked drop in performance, they are taken out of service for a complete overhaul during which any parts worn to certain limits will be replaced. This is a useful wet season activity for the workshop. Preventive maintenance makes parts stock planning easier because dates can be fixed and ample lead times given for the purchase of clearly defined selections of parts.

Irrigation pump engines worth noting are often equipped with safety devices which stop the engine automatically if the oil pressure drops or coolant temperature rises above certain limits. The usual system employs a solenoid to hold the fuel pump stop control open as long as the engine is running and current is flowing to the solenoid. An oil pressure gauge and a coolant temperature gauge are wired into the system so that the indicator needles will touch electrical contacts on the faces of the gauges if low pressure or high temperature conditions arise, thus switching current from the solenoid and allowing the pump control to close. This is a fail-safe system. Consequently, if any part of it is defective the engine will refuse to run. Parts such as the special gauges and solenoids should be stocked. If such parts are not available there will be a tendency for the operator to eliminate the safety system and run the engine without it.

Similar safety systems are often used on generator sets and it is equally important that they should be kept in working order.

6.20 Grain handling and feed preparation machinery

While parts for engines, tractors and trucks are commonly stocked in some sort of dealers store, parts for grain and feed machinery are seldom treated in this way. Grain drying and storage and feed milling installations are often tailor-made to specific situations; parts are more specialized and less easily available.

Nevertheless, there are agricultural machinery dealers who service grain and feed machinery and stock parts for it. In a project, remote from supply sources, there is no reason why parts for this equipment need be treated any differently from tractor or truck parts.

Increasingly, grain and feed machinery is driven by electric motors. Sometimes one motor drives many machines through a line shaft running a number of belt drives. In this case, stocks of belts, belt fasteners and belt treatment (anti-slip) compounds will be needed. More often, however, there will be individual motors for each machine. Motor controls may be individual or linked and integrated into a central switchboard. Replacement motors may be needed but this should be infrequent if the control gear and its protective devices are kept in good condition.

Central control panels will require a stock of relays, overload protection devices and possibly timers and indicator lamps. Installations containing electronic programmes may need replacement printed circuit boards since these can deteriorate under operating conditions where heat and humidity cause corrosion.

Each machine in the installation will have some unique parts requirements. Hammer mills will need hammers. Rice mills will need rubber roller covers and abrasive materials for the polishers. Grain drying furnaces will need burner parts. Such parts will usually be few in number and can be clearly identified by the manufacturer.

The majority of parts required by grain and feed plants will be more common items. Drive components such as belts, chains, sprockets, pulleys, bearings, keys and shafts are likely to be standard specification items and available from a variety of industrial supply companies. Hardware will be a frequent requirement and will largely be the same as that stocked for mobile equipment.

6.21 Construction equipment and forklifts

Parts stocking for construction equipment is worth a separate publication given the wide variety of machines and their large and expensive nature. The economics of machinery selection and use are very precise which implies a parallel degree of precision in the selection of parts stock. However, there are similarities between construction machines and agricultural equipment. Maintenance parts are equally important. Some consumables and much of the hardware is the same. A greater range of hardware will be needed including larger bolts in higher tensile strengths. Engine parts will be similar but there are some idiosyncrasies, particularly in fuel systems.

Tracked machines require special attention since tracks contain many moving parts which, under abrasive soil conditions, will be subject to rapid wear. Some of these, particularly the flanges on track rollers and the track links themselves, can be re-built. In some cases sprockets can be fitted with a new tooth ring instead of replacing the whole sprocket. These points will be clear from manufacturers' repair manuals. Seals will be needed, both for the rollers and for final drive housings and pins and bushes will be needed for the tracks in addition to the special track bolts and nuts.

Other important parts for construction machinery are the ground-engaging parts such as teeth and cutting edges which need to be replaced from time to time, and the hydraulic system which is subject to a great deal of wear and tear. Some operate at high pressures and some use large volumes of oil which is usually expensive and not always available off the shelf. Stocking plans may therefore have to include special hydraulic oils (also transmission oils) and hoses, couplings, end fittings, ram seals and pump components. Complete pumps may be stocked as service exchange units.

Parts Order Form   Appendix A

Name of supplying organization _____						
Address						
From (buyer) _____				Order No. _____ Date _____ Dispatch by _____ Mark _____			
Send to						
Terms of payment _____ Account No. _____				Invoice No. _____ Date _____			
Letter of credit No. _____ Import licence No. _____				Expiry date _____ Expiry date _____			
Item	Description	Part No.	Qty	Price	Locn.	BO
Signed _____ Date _____						
Page _____ of _____						

Parts Invoice

Name (Of organization invoicing the parts and using pre-printed form sets of which both the order and the invoice are parts).
Address

To: (name of buyer)	Order No. _____
	Date _____
	Dispatch by _____
Address:	Mark:
Send to:	
Terms of Payment:	Invoice No. _____
Account No. _____	Date _____
Dispatched by:	Date _____
Carrier:	Way Bill No. _____

Item	Description	Part No.	Qty.	Price	Total	Bo
				Total Parts		
Signed:			+ Carriage		
Date:			− Discount	_____	
			TOTAL		

Page ____ of ____

Modern, International Export Invoice

Seller's name and address	Invoice No. _____ Date___ Order No. _____ Other references _____		
Consignee's name and address	Buyer (if not consignee)		
	Presenting bank		
Port of Lading	Country of origin		
Country of final destination	Ship/Air/Etc.	Terms of Deliv. & Payment	
Other transport information	Currency of Sale		
Marks & Numbers	Description of Goods	Gross Wt.	Net Wt.	Vol. m^3
No. and kind of packages	Qty	Unit price	Total
CERTIFICATE OF VALUE AND ORIGIN We hereby certify that the goods in this invoice have been produced wholly/ substantially in _____ and that this invoice shows the actual price of the goods and all particulary are true and correct in accordance with our books. Signed _____	Total goods Packing Carriage Port Charges Total FOB Freight Insurance Other Charges TOTAL		

Forwarder's Air Waybill

Forwarder's name Address			House Airbill No.		
Airport of Departure	Routing		Destination Airport		
Consignee's Name & Address	Also Notify				
	Flight		Day		Service
Shipper's Name & Address	Multiple Air Waybill No.				
Shipper's Order No.	Account No.		Customs Reference No.		
CERTIFICATION: WE HEARBY DECLARE that the goods mentioned herein were dispatched by .... on .... by flight .... Carrier certifies goods were received for carriage subject to the terms and conditions of the Institute of Freight Forwarders Ltd. Signed.............as agent for issuing carrier.					
Number of Packages	Actual Gross Weight Kilos	Nature and Descript.	Chargeable Weight	Rate	Total
Dimension Marks and Numbers (cm)			Letter of Credit Details Other Information		
Currency	Shipper's Declared Value for		Docs. to Accompany Shpt.		
	Carriage	Customs	Insurance	Consular Inv.	Commer. Inv.	Certif. of Orig
Invoice Details Prepaid Freight (see above) Inbound Freight Insurance Local Carriage Local Processing Disbursements C.O.D. TOTAL			xxxxx xxxxx	Collect	xxxxx

Parts Picking Document

```
Name (Of organization invoicing the parts and using
pre-printed form sets of which both the order and
the invoice are parts).

Address


To: (name of buyer)           Order No. _____
                              Date _____
                              Dispatch by _____
Address:                      Mark:



Send to:



Terms of Payment:             Invoice No. _____
Account No. _____        Date _____

Letter of Credit No. _____  Expiry Date _____
Import Licence No. _____  Expiry Date _____
```

Item	Description	Part No.	Qty.	Picked	Locn.

```
Signed:
Date:
     Page ____ of _____
```

Stock Report

Date	Opening Stock	+Cost of Parts in	-Cost of Parts out	Closing Stock	Month Sales	Planned Stock

Movement Report

Date	Month Sales	Total YTD	Plan YTD	Var. %	Ords No.	Items No.	Ord $	Item $	Item/ Ord

Service Level Report

A. Items Supplied		B. Not Supplied				C. Service		
		No.		Value				
No.	Value	S.	N.S.	S.	N.S.	No.	Value	Plan Service Level

Back Order Report

Back Orders Supplied		% of Total Items	
No. of Items	Value	No.	Value

Trading Summary Report

Sales		Gross Profit			Expenses			Net Profit		
Value	Cost	M	YTD	Plan	M	YTD	Plan	M	YTD	Plan

Stock Replacement Indent Form

Stock Type						Lead Time Mo.			Date		
Desc.	Part No.	Sales				Ord Lvl	Act Bal	Ord Qty	To Order		
									Qty	Price	$
		Q1	Q2	Q3	Q4	Av					

Inventory Record Card

Order Card

Supplier 1:								Address:																					
Supplier 2:								Address:																						
Supplier 3:								Address:																						
Date	Reference	Sup	Ord	Recd. Qty	Recd. Dte	Cost	Recd. Qty	Recd. Dte	Cost	Recd. Qty	Recd. Dte	Cost	Bal	Min
Replaces				Replaced By				Before S. No.			After S. No.			
Part Number				Name				Vehicle			Assembly			Location

Movement Card

Date	Reference	In	Out	Balance	In	Out	Balance	Replaces	Replaced By	Before S. No.	After S. No.	Part Number	Name	Vehicle	Assembly	Location

Appendix B

Personnel Application Form

Company name Address Vacancy_____   Ref. _____
Application from

Family name_____   Other names_____
Address   Telephone no. _____
                           Married____ Single ____
                           Children (ages) _____
Date of birth _____   Age ____ Nationality_____

Degrees and/or technical qualifications   Driving Licence? ___   Criminal convictions? _____
Schools attended (give dates and certificates)
Present position_____   From_____ Salary_____ Name of employer _____ Address   Telephone no. _____   Brief details of duties Name of supervisor who may be contacted for reference Please indicate if you do not wish present employer to be contacted. _____ Signed _____   Date _____

Employment Record
(Reverse of Personnel Application Form)

Employer _____ From _____ To _____
Address   Telephone _____
Position _____ Salary _____
Reason for leaving
Employer _____ From _____ To _____
Address   Telephone _____
Position _____ Salary _____
Reason for leaving
Employer _____ From _____ To _____
Address   Telephone _____
Position _____ Salary _____
Reason for leaving

Structured Interview

Interview by _____ Date _____	
By telephone? ____ In person? _____	
Name of reference _____ Position _____	
Address _____ Telephone _____	

Name of applicant _____
Present position _____ From _____
Duties

Competence

Reliability

Timekeeping Good? ____ Poor? ____
Absenteeism Yes? ____   No? ____

Qualifications (check applicant information)

Education (check applicant information)

Employment record (Check application information)

Recommendation based on this interview

Signed _____

Employee Record

Name _____	Date of birth _____
Address	Telephone _____
	Starting date _____

Position _____ Salary _____

Names of dependents and addresses

Year _____
Position changes   Salary changes

Training courses   Graded

Supervisor's reports (dates)
Warnings: verbal _____ written _____
Progress: satisfactory _____ unsatisfactory _____
Recommendations

Signed _____ Date _____

Year _____
Position changes   Salary changes

Training courses   Graded

Supervisor's reports (dates)
Warnings: verbal _____ written _____
Progress: satisfactory _____ unsatisfactory _____
Recommendations

Signed _____ Date _____

Appendix C

Suggested Training Programme

Course I. <u>Basic Parts Skills.</u>

Target: all new staff;
existing junior staff as a refresher.

Time required: one hour per day for four days.

Syllabus: a. Introduction - background to parts requirements, objectives, definitions and situations;

b. Basic engineering principles reflected in the parts store;

c. Basic storekeeping techniques;

d. Details of policy regarding operational procedures of the organization.

Course II. <u>Warehouse Procedures.</u>

Target: warehouse staff;
other junior staff, to broaden their knowledge.

Time required: one hour per week for two weeks.

Syllabus: a. Receiving and binning procedures, neatness and order;

b. Issue procedures, stock movements, inventory checks and security.

Course III. <u>Administrative procedures.</u>

Target: clerical staff;
other staff, to broaden their knowledge.

Time required: one hour per week for six weeks.

Syllabus: a. Procedures for processing incoming goods including checking, document routing, payments and entry on cards;

b. Procedures for processing outgoing parts including customer orders, withdrawal documents and invoices;

c. Back orders - supply and customer;

d. Order compilation including order levels, order points and Economical Order Quantity;

e. Stock record cards - uses, entries, renewal, substitution;

f. Costing and pricing;

g. Filing, correspondence and other office procedures.

Course IV. <u>Parts Sales.</u>

Target: sales staff;
other staff to broaden their outlook.

Time required: one hour per week for four weeks.

Syllabus: a. Parts identification;

b. Sales procedures - customer orders, processing, invoicing and back orders, credit control;

c. Customer relations- sales techniques, problem solving.

Course V. <u>Inventory Control.</u>

Target: supervisors;
stock record clerks.

Time required: one hour per week for two weeks.

Syllabus: a. Inventory control theory;

　　　　　b. Inventory control as practiced by the
　　　　　　organization.

Course VI. <u>Equipment Update.</u>

Target: all staff.

Time required: one hour, four to six times per year.

Syllabus: a. Review of modifications to existing
　　　　　　equipment;

　　　　　b. Population changes;

　　　　　c. New models.

Course VII. <u>Management.</u>

Target: supervisors.

Time required: one hour, four times per year.

Syllabus: a. Communication;

　　　　　b. Motivation;

　　　　　c. Discipline;

　　　　　d. Planning and control.